What People Are Sa

Am I Too Old to Save

This small book is a gem of great value: it will turn more of us older Americans from complainers and worriers into people who are fixing the problems that assail us. It makes a powerful case for generational action; if the kids can do it so can we!
Bill McKibben, founder of Th!rdAct and author of *The Flag, the Cross and the Station Wagon*

All my hand-wringing progressive friends need to read *Am I Too Old to Save the Planet?* now. MacDonald's timely handbook on the climate crisis not only offers concrete steps we can take as individuals to transform the future but, importantly, invites us to join with others in collective action to bring about the systemic change that is needed. It's about we, not me.
Helen Stoltzfus, playwright, performer, Co-Artistic Director, Black Swan Arts & Media

OK, Boomer, here's your chance to prove a thousand memes wrong. Lawrence MacDonald has written the ultimate, clear-eyed guide to make a difference, and it goes well beyond refusing the plastic bag at the checkout counter and other single actions. It's about feasibility, impact, and spreadability, as the book states clearly. Just do it indeed!
Gernot Wagner, climate economist, Columbia Business School, and author of *But Will the Planet Notice? How Smart Economics Can Save the World*

As a member of the baby boom generation, I'm well aware of the skills and connections we acquire over our lives that can be used to make the world a better place. MacDonald's book

provides a practical guide for people of a "certain age" who want to leave a livable planet for future generations. We can make a difference at any age. This book details the many things that boomers have to offer.
Madeleine Para, Executive Director, Citizens' Climate Lobby

Drawing on his experience as a climate advocate and a career in policy communications, MacDonald spells out in clear and actionable language what each of us can do, on our own and with others, to confront the existential challenge of our time. This is a how-to guide for those who want to do their part to stand up to the climate crisis. It's essential reading for those who want to look our children in the eye - and leave them a livable future. It's a message of hope grounded in the timeless truth that it's never too late to make a difference.
Manish Bapna, President and CEO of the Natural Resources Defense Council

This thoughtful and highly readable book is an honest look at a powerful (and often under-appreciated!) generation and all they can still do to save our collective future. Lawrence MacDonald sounds an urgent call and offers concrete ways to take meaningful climate action. A critical book for our times.
Rabbi Jennie Rosenn, founder and CEO of Dayenu: A Jewish Call to Climate Action

I read many climate books, and I found fresh insights and useful reminders in this book. Lawrence MacDonald delivers a breezy walk through what people can do individually and, most importantly, collectively. *Am I Too Old to Save the Planet?* offers a handy synthesis for people at all stages of their climate learning and action journey.
Greg Dalton, founder and host of *Climate One Podcast*

This book is a powerful call to baby boomers to protect our world and our legacy from the calamity of climate change. It is not too late to ensure that our children and grandchildren inherit a healthy planet and enjoy a good life. This book provides a road map for reaching that goal.
Mona Sarfaty, founder of Medical Society Consortium on Climate and Health, Center for Climate Change Communication, George Mason University

Lawrence MacDonald's *Am I Too Old to Save the Planet?* answers strongly: No! We elders are living longer and healthier to take responsibility for this essential role. We were born for this, and now is our time to act as good ancestors! MacDonald's masterpiece shows us the way.
John Sorensen, founder of Elders Action Network

In tracing his own arc of awareness of the need for increased climate activism, Lawrence gives baby boomers a thoughtful and inspiring guide to up our game. Third Act's mottos of 'no time to waste' and 'old and bold' are writ large on every page.
Deborah Kushner, co-founder of Th!rdAct Virginia

OK boomers, you're up! Lawrence MacDonald has given his generation a warm invitation and a road map (remember those?) for bringing their much-needed power and experience to the climate movement. Whoever we are, MacDonald reminds us, we are already exactly who we need to be to respond to the climate crisis with integrity and compassion. To anyone at any age who is ready to find your place among the climate protectors fighting for the common home we share, welcome — this book is for you.
Joelle Novey, Director, Interfaith Power and Light, DC, Maryland, and Northern Virginia

As a late boomer myself, I couldn't put this book down. MacDonald brings moral clarity and practical know-how to the urgent premise: Boomers can and should do more to fight climate change. Be careful, this book could change your life.
Mike Tidwell, Director of Chesapeake Climate Action Network

Am I Too Old to Save the Planet?

A Boomer's Guide to Climate Action

For Muriel and Isaac

Am I Too Old to Save the Planet?

A Boomer's Guide to Climate Action

By Lawrence MacDonald

Winchester, UK
Washington DC, USA

JOHN HUNT PUBLISHING

First published by Changemakers Books, 2023
Changemakers Books is an imprint of John Hunt Publishing Ltd., No. 3 East Street,
Alresford, Hampshire SO24 9EE, UK
office@jhpbooks.com
www.johnhuntpublishing.com
www.changemakers-books.com

For distributor details and how to order please visit the 'Ordering' section on our website.

ISBN: 978 1 80341 484 3
978 1 80341 485 0 (ebook)
Library of Congress Control Number: 2023935076

A CIP catalogue record for this book is available from the British Library.

Design: Lapiz Digital Services

UK: Printed and bound by CPI Group (UK) Ltd, Croydon, CR0 4YY
Printed in North America by CPI GPS partners

The Resetting Our Future Series

At this critical post-pandemic moment in history, we have the rare opportunity for a Great Reset – to choose a different future. This series provides a platform for pragmatic thought leaders to share their vision for change based on their deep expertise. For communities and nations struggling to cope with the crisis, these books will provide a burst of hope and energy to help us take the first difficult steps towards a better future.
—Tim Ward, publisher, Changemakers Books

Am I Too Old to Save the Planet? A Boomer's Guide to Climate Action by Lawrence MacDonald, Th!rd Act Climate Activist and former Vice President, World Resources Institute

What if Solving the Climate Crisis Is Simple? by Tom Bowman, President of Bowman Change, Inc., and writing-team lead for the U.S. ACE National Strategic Planning Framework

Zero Waste Living, the 80/20 Way: The Busy Person's Guide to a Lighter Footprint by Stephanie Miller, founder of Zero Waste, in DC, and former Director, IFC Climate Business Department

A Chicken Can't Lay a Duck Egg: How COVID-19 Can Solve the Climate Crisis by Graeme Maxton (former Secretary-General of the Club of Rome) and Bernice Maxton-Lee (former Director, Jane Goodall Institute)

A Global Playbook for the Next Pandemic by Anne Kabagambe, former World Bank Executive Director

Power Switch: How We Can Reverse Extreme Inequality by Paul O'Brien, Executive Director, Amnesty International USA

Impact ED: How Community College Entrepreneurship Creates Equity and Prosperity by Rebecca Corbin (President & CEO, National Association of Community College Entrepreneurship), Andrew Gold and Mary Beth Kerly (both business faculty, Hillsborough Community College)

Empowering Climate Action in the United States by Tom Bowman (President of Bowman Change, Inc.) and Deb Morrison (Learning Scientist, University of Washington)

Learning from Tomorrow: Using Strategic Foresight to Prepare for the Next Big Disruption by Bart Édes, former North American Representative, Asian Development Bank

Cut Super Climate Pollutants, Now! The Ozone Treaty's Urgent Lessons for Speeding Up Climate Action by Alan Miller (former World Bank representative for global climate negotiations), Durwood Zaelke (President and founder, the Institute for Governance & Sustainable Development) and Stephen O. Andersen (former Director of Strategic Climate Projects at the Environmental Protection Agency)

Resetting Our Future: Long Haul COVID: A Survivor's Guide: Transform Your Pain & Find Your Way Forward by Dr. Joseph J. Trunzo (Professor of Psychology and Department Chair at Bryant University), and Julie Luongo (author of *The Hard Way*).

SMART Futures for a Flourishing World: A Paradigm Shift for Achieving Global Sustainability by Dr. Claire Nelson, Chief Visionary Officer and Lead Futurist, The Futures Forum

Rebalance: How Women Lead, Parent, Partner and Thrive by Monica Brand, Lisa Neuberger & Wendy Teleki

Provocateurs not Philanthropists: Turning Good Intentions into Global Impact by Maiden R. Manzanal-Frank, President and CEO, GlobalStakes Consulting

Resetting the Table by Nicole Civita (Vice President of Strategic Initiatives at Sterling College, Ethics Transformation in Food Systems) and Michelle Auerbach (Educator and founder of Modaka Communications)

Unquenchable Thirst: How Water Rules the World and How Humans Rule Water by Luke Wilson and Alexandra Campbell-Ferrari (Co-Founders of the Center for Water Security and Cooperation)
www.ResettingOurFuture.com

Contents

Preface

"You don't give a damn about us," Greta Thunberg, the youth climate activist and founder of the School Strike for Climate, has repeatedly told powerful adults. "The symbolism of the school strike is that since you adults don't give a damn about my future, I won't either."

Chole Swarbrick, a 25-year-old New Zealand lawmaker, caught global attention in 2019 when she casually dissed an older member of parliament who heckled her while she was giving a speech calling for a national zero emissions target. "OK boomer," she deadpanned, continuing her speech without missing a beat.

Young people like Thunberg, Swarbrick and millions of others around the world are at the forefront of the climate justice movement. Where are we, their elders, who should be actively and visibly supporting them? I wrote this book because I believe that it's morally unacceptable for the generation that did the most to create the problem to sit back and leave it to the kids to deal with it.

Fortunately, many of us do give a damn. We care about the planet, about national and global poverty, about rising inequality, and about the looming mass extinction. We want to pass along a livable world to those who come after us. We want our children and grandchildren—and all future generations—to have clean air and water, predictable seasonal cycles, and access to an outdoors filled with natural wonder.

If you are boomer who gives a damn but isn't sure how you can make a difference, this book is for you.

There are millions of us. A 2021 Pew Research study found that 57% of Americans who are boomers and older say that climate should be a top priority for assuring a sustainable planet for future generations. The Yale Program on Climate

Communications found that 10% of boomers said that they would definitely or probably participate in non-violent civil disobedience such as sit-ins, blockades or trespassing to block activities that make global warming worse. Fully two million said that they would "definitely" do so if asked by somebody they liked and respected.

Properly mobilized, boomers can make a world of difference. Our generation holds 70% of U.S. household wealth. We wield immense power and influence in institutions such as government, corporate boards, alumni associations and religious organizations. And we vote. What we haven't done—yet—is to organize as a generation around climate change.

Imagine if we boomers who are not OK with leaving future generations an inhospitable planet began acting now, in the time that remains to us, as if the world depended on what we do. Imagine if we resolved to do all we can to address the climate emergency. Imagine if we started with our personal actions, like eating less meat and installing rooftop solar, and went on to organize our generational power to hasten system change, helping young people to avert climate catastrophe.

Imagine being able to say to Greta Thunberg and the angry, grieving members of her generation "Yes, some of us do give a damn. We are sorry our generation didn't act sooner. We will do all we can, starting now, to avert climate catastrophe."

This book is addressed to the majority of boomers who recognize that climate change is an urgent threat and understand the need for a rapid, systemic transformation from obscenely unequal, low-efficiency, take-make-waste, high-carbon economy to a more egalitarian, high-efficiency, circular and low-carbon economy.

The oldest of us is now 77, the youngest 59. Given increased longevity, many of us can reasonably expect at least another decade or more of healthy, active living. And while rising inequality and the slipshod American safety net mean that millions of us must worry about rent, food, and medical expenses, millions of others find ourselves surprisingly comfortable, enjoying reasonably good health and the benefits of tax-free retirement savings and surging home values.

How shall we spend our time in retirement? Sports? Second homes? Motorcycle riding? Exotic vacations? The approximately 10 years of active, healthy life that remain to many of us coincide with the last 10 years in which humans can avoid climate catastrophe. Shouldn't those of us who understand the risks and have the means be joining our children and grandchildren in demanding climate action? Is there anything more important we should be doing with our time?

Although there are dozens of books about boomers and hundreds of books about climate change, this is the first to examine the U.S. response to climate change through a generational and global lens. Part I shows how boomers shaped a one-step-forward-two-steps back U.S. policy and how the resulting lack of U.S. leadership undermined the global response. Part II offers a boomer's guide to action that combines history, research on the climate activism landscape, my own experiences, and interviews with boomer activists.

Because this is a boomer's guide—specifically my guide—it's perhaps appropriate to share a little about me. I was born in 1954, the middle of the post-World War II baby boom. Although I was too young to join fully in the seminal events of the late 1960s I, like others in my generation, shared in the music, language and protest culture of the era. After graduating from the University of California at Santa Barbara, I went to Taiwan to study Chinese and spent much of the next 17 years in Asia, mostly working as a reporter. Returning to the United States in 1993, I worked as a policy communications professional at the World Bank and two DC-based think tanks.

Along the way, I observed the outsized role of the United States in shaping world affairs and the outsized role of my generation in shaping U.S. policy. At the World Bank, the Center for Global Development (CGD), and in my most recent job as the vice president for communications at the World Resources Institute, I devoted the second half of my career to communicating science-based solutions to the world's most pressing problems, especially climate change.

Gradually, I recognized that problems were getting worse not because of a lack of technical solutions but because of the power of those who benefit from the status quo to block needed changes, and the lack of power of ordinary people who want an equitable, livable planet to overcome them. I decided when I retired last year to become a full-time climate activist and to write this book to invite others of our generation to the join the growing ranks of climate boomers.

As I finish this book, I am sharply aware of its limitations, and my own. Climate change is a profound injustice that the rich and powerful, mostly white men in the global North, are inflicting on others, mostly poorer people of color, all over the world. As a white male boomer, I drew a lucky ticket in life and have benefitted in ways that I am only beginning to understand. In telling the story of my generation and the climate crisis, I've tried to overcome the resulting myopia. In the end, however, this remains a story written by a white male boomer. For these and other shortcomings in this book, I take full responsibility.

But enough about me. I want to acknowledge three people whose thinking and writing greatly influenced mine and thus helped to shape this book.

Development economist Nancy Birdsall, the founding president of CGD, taught me and many others how to examine the policies of the United States and other rich countries for their impact on the lives of the 6.7 billion people who live in developing countries. I am grateful to her as a role model and a friend.

Rebecca Solnit's work has inspired me since my daughter handed me the third edition of *Hope in the Dark: Untold Histories, Wild Possibilities* (2016). Solnit's *Orwell's Roses* (2019), emboldened me to try weaving together reporting, accounts of other books, and personal experience. I'm eagerly awaiting publication of *Not Too Late: Changing the Climate Story from Despair to Possibility*, a volume of essays that Solnit edited with Thelma Young Lutunatabua.

Bill McKibben's thought leadership has been so central to the development of my own views that I no longer remember whether his 2006 essay calling on boomers to become climate activists planted the idea in my head or merely articulated something I already believed. I am grateful to him for the kind words he has written about this book and I'm proud to be a member of the growing tribe of experienced Americans he is mobilizing through Th!rdAct to fight with love and creativity for a livable planet.

Of the many other friends who encouraged and advised me, three stand out.

David Hindin, a former enforcement attorney and executive at the Environmental Protection Agency, has been my companion on an intellectual journey that has led us both to the conclusion that moderate, incremental approaches to climate emergency are not enough. His perceptive comments on the near-final draft improved the book.

Howard Smith brought a fresh perspective and designer's eye to the topic. He suggested the title—*Am I Too Old to Save the Planet?*—prompting me to adopt a more conversational tone that I hope saves the book from being too preachy. He also created the cover and suggested and designed the part and chapter dividers, signposts that I hope make the line of argument easier to follow. Smith was one of several reviewers who urged me to add the checklists at the end of each chapter. I hope you find them useful!

This book would not exist without the encouragement and support of Tim Ward, who oversees the Changemakers Books imprint at John Hunt Publishing. Ward was the first person to encourage me to write this book, and his coaching and advice have been pitch-perfect throughout the journey. I am proud to call him my publisher and my friend.

As I neared the finish line, my millennial, climate activist adult children, Muriel and Isaac, helped me see the book through their generation's eyes, offering valuable suggestions that greatly improved it. I am fortunate indeed that my patient and loving wife, Hannah Moore, is also an eagle-eyed proof reader. In this project, as in life, she saved me from embarrassing errors.

Finally, although this book is addressed to fellow boomers, I wrote it not for us but for our children and our children's children, and for the many generations that will follow. I hope that when these future generations look back at our era, they will know that a large number of boomers did everything we could in the time that remained to us to bequeath to them a livable planet.

Part 1

How we got here

Chapter 1

Are boomers to blame?

Climate change went from a manageable problem to a planetary emergency on our watch. We boomers are overdue for a reckoning with our collective responsibility.

Chapter 1

Are boomers to blame?

America's boomers—those of us born between 1946 and 1964—are the generation most responsible for the climate emergency. We also have the power, in the time that remains to us, to play a key role in averting a planetary catastrophe that otherwise will last for millions of years.

This book is for boomers. If you, like me, are an American baby boomer, it will show you how to reconnect with the best of your youthful ideals and join with others to leave a lasting, positive legacy to the generations that come after us.

We have a lot to make up for. American boomers benefitted massively from the post-World War II fossil fuel-driven economic boom, riding it to unprecedented prosperity. We not only emitted greenhouse gases by driving and flying—although, to be sure, we did plenty of that. We also lived and worked in a world of material abundance fueled by coal, oil and methane. This was our advantage, and it has come at tremendous cost. We will leave behind us the biggest carbon footprint of any generation in history.

Although we learned about the climate threat while there was still time to manage a smooth transition, we failed to heed ever louder and more urgent alarms. We didn't just fail to make lifestyle choices that would have cut our own emissions. We also failed to demand that U.S. politicians enact policies to avert this planetary crisis.

Our failure as a generation has had global repercussions. As the richest and most powerful nation that the world has ever seen—and the country that has emitted the most heat-trapping gasses—the United States has the means and the obligation to lead a global response to the climate threat. The Americans before us overcame the Depression, defeated fascism and, after World War II, established the United Nations, the World Bank, the International Monetary Fund and other global institutions to prevent a third world war. But in the face of the greatest threat that humanity has ever faced, we boomers have done next to nothing.

We didn't lack for high ideals. Millions of us participated in peaceful protests that ended the war in Vietnam and joined in the first Earth Day, earning a reputation as America's "First Green Generation." As we grew older, however, we became distracted and self-indulgent. Instead of leading the way to a clean, low-carbon economy, America's boomers—and the boomer politicians we elected—took one step forward, two steps back, stalling America's energy transition and sending mixed signals to the world that undercut the fragile global consensus for action.

Boomers at Center Stage

We boomers have been center stage throughout America's on-again, off-again efforts to address the climate challenge. A fellow boomer, Vice President Al Gore, helped negotiate the 1997 Kyoto Protocol, the first international treaty to cut greenhouse gasses. Nearly every nation—191 countries—ratified it. Only the United States failed to do so. Bill Clinton, our first boomer president, signed it but never sent it to the boomer-dominated Congress, where it lacked the votes for ratification. In 2010 the United States quietly withdrew.

The story of the 2015 Paris Agreement negotiated under President Obama is more complex but equally shameful. Negotiators representing the nations of the world, informed by U.S. diplomats that the United States would never accept anything binding, designed the accord to be toothless. To avoid the need for U.S. Senate ratification, the Paris Agreement is not technically a treaty and lacks any enforcement mechanisms.

A late boomer born in 1961, President Obama had campaigned on a promise of hope and change that included a commitment to address climate change. When he clinched the Democratic nomination in 2008, he famously predicted that people would look back and say: "this was the moment when the rise of the oceans began to slow and our planet began to heal."[1] At the 2014 UN Climate Change Summit he urged action, saying: "we are the first generation to feel the impact of climate change and the last generation that can do something about it."[2]

When the Paris Agreement hung in the balance at the 2015 climate conference, Obama phoned Chinese leader Xi Jinping in a last-ditch effort to bring the negotiations to a successful conclusion. He and Xi later signed the Paris Agreement on the same day, a symbolic commitment by the world's two biggest emitters to work together to address the climate crisis. With deterioration of U.S.-Chinese relations since then, that may have been a highwater mark for such cooperation.

But Obama's soaring rhetoric and personal diplomacy notwithstanding, he deployed neither his executive authority nor the bully pulpit of his presidency to accelerate the end of the fossil fuel era. Worse, he adopted an "all-of-the-above" energy policy that boosted federal subsidies for fossil fuel extraction. In Obama's second term, he lifted a ban on crude oil exports and authorized for the first time the export of methane gas, production of which had surged on his watch due to a government-subsidized fracking boom.[3] Two years after he left office in 2016, the United States surpassed Saudi Arabia and Russia to become the world's largest producer of crude oil.

By then the Paris Agreement had become a political football. Despite the diplomatic contortions that the international community had undertaken to accommodate the United States, Donald Trump—an early boomer born in 1946 and likely America's last boomer president—withdrew from the agreement, making the United States the only nation in the world to drop out.

Joe Biden—born in 1942 and almost certainly the last pre-boomer president—kept a campaign promise to re-join the Paris Agreement on his first day in office. But the Senate's failure to pass his Build Back Better climate and social welfare bill before the U.N. climate summit in the fall of 2021 meant that U.S. negotiators went empty-handed to Glasgow, Scotland, for the first UN climate conference of the Biden administration. Biden's efforts to reassert American leadership on climate rang hollow.

The United States was in a somewhat stronger position at the UN climate conference in 2022 following the passage of the so-called Inflation Reduction Act, a scaled down version of Build Back Better that was nonetheless the most significant climate action legislation in U.S. history.

But the conference itself was deeply flawed. Held in the Egyptian coastal resort Sharm el-Sheikh and sponsored by Coca-Cola, one of the world's top plastic polluters, it included more than 600 representatives from fossil fuel companies, among them several CEOs. Although negotiators eventually agreed to set up a "loss and damage" fund to compensate countries hard-hit by climate disasters, there was no plan for how this would work or where the money would come from.

Alok Sharma, the British politician who had overseen the UN conference in Glasgow, was frustrated that the fossil fuel companies and countries that have been slow to cut emissions colluded to weaken the text that emerged from the conference. "Peaking emissions by 2025 is not in this text. Follow-through on the phasedown of coal is not in this text. The phasedown of all fossil fuels is not in this text," he said.

Without a dramatic reduction in emissions of heat-trapping gasses—primarily CO_2 and methane from fossil fuels—the world will soon blow past the Paris Agreement stretch goal of holding average temperature increase by the end of the century to 1.5 degrees Celsius (2.7 degrees Fahrenheit) above the pre-industrial average. Worse, a growing number of scientists warn that global average temperatures may surpass the Paris Agreement's fallback goal of 2 degrees Celsius (3.6 degrees Fahrenheit), resulting in rapid, extreme sea-level rise and recurring heat waves that would make much of the world where people currently live uninhabitable.

"We are on a highway to climate hell with our foot on the accelerator," UN Secretary General António Guterres told world leaders at the opening of the climate conference in Sharm el-Sheikh. "The global climate fight will be won or lost in this crucial decade—on our watch. One thing is certain: those that give up are sure to lose."

Speaking again at the end of the meeting, he said: "a fund for loss and damage is essential—but it's not an answer if the climate crisis washes a small island state off the map or turns an entire African country to desert. The world still needs a giant leap on climate ambition."

A century from now, when future historians try to make sense of humanity's response to the climate threat, it will be clear that American boomers held the steering wheel at a critical moment. We, the generation that made giant SUVs the country's best-selling vehicle, are now speeding toward a well-marked interchange where humanity can either continue headlong into a planetary pileup or veer away at the last minute onto an exit that just might lead us to preserve a livable planet. Which shall it be?

So How Is This My Fault?

At this point you may be wondering, "so how is this my fault?" Of course, no single individual can be asked to shoulder the responsibilities of an entire generation. But because climate change is fundamentally a question of intergenerational injustice—what our generation does and fails to do will impact countless generations yet unborn—I believe that we boomers are overdue for a reckoning with our collective responsibility.

How does collective responsibility relate to individual responsibility? You have likely heard the quote, often attributed to the philosopher Edmund Burke, that "all that is necessary for the triumph of evil is that good men do nothing."[4] Although Burke did not write these precise words, the quote persists because it captures something we know to be true: bad things happen when good people fail to stop injustice. Whether we actively abetted the climate crisis or merely did nothing much to stop it, each boomer owns a piece of this planetary catastrophe, because it happened on our watch.

Reckoning with our generation's role—and our personal share in it—can be extremely painful. In conversations with boomer friends while writing this book, I have found that many of them know all too well that we are in danger of leaving future generations an unlivable planet. They feel terrible about it, but also powerless, so they don't want to think about it. "Don't make me feel guilty!" is a refrain I have heard again and again.

The problem is, if we try to ignore what we know to be true, we are living in a state of denial. We can only free ourselves from our climate culpability by acknowledging it and vowing to do all that we can to repair or minimize the damage. Crucially, knowing that there are things we can do that will make a real difference makes it possible for us to face up to our responsibilities.

This approach is based on ancient wisdom. Many faith traditions teach that acknowledging our mistakes—what religious folks call sins—involves two processes: repentance and atonement. Repentance requires recognizing and confessing the errors of our ways and vowing not to repeat them. Atonement requires doing all we can to set things right. The Twelve Steps of Alcoholics Anonymous embodies this same idea: acknowledging the harm one has caused and making amends where possible are seen as crucial steps towards recovery.

Acknowledging our climate culpability is intellectual and emotional work that each of us must undertake individually and collectively. The next chapter, "What Went Wrong?" provides a guide, recalling how our generation began with high hopes and ideals but failed to live up to them when it came to the crucial job of protecting the planet. First, however, we must understand the severity of the climate threat which, for reasons we will explain, is even worse than what you probably think.

How Bad Is It?

An accurate understanding of the urgency and scope of the problem is fundamental to deciding what to do about it. While much of what I'm going to suggest in the chapters that follow is non-controversial, I conclude by exploring the need for the climate movement to develop a strong radical flank and suggesting ways that boomers can help. Some of these suggestions may go beyond what you expect. These militant actions must be considered in the context of the incredible risks of continued inaction.

So, how bad is it? Unless you are a climate scientist or a policy wonk who tracks climate impacts for a living, climate change is almost certainly worse than you think. Even for those who closely follow mainstream media climate news, the magnitude of what's happening is not easy to grasp.

That's partly because climate change is so vast, extending across all the globe and infinitely into the future. Borrowing from computer programming jargon, environmental philosopher Timothy Morton coined the term "hyperobject" to describe climate change and other phenomena that are "massively distributed in time and space" and therefore very hard for humans to comprehend. Other examples of hyperobjects, he writes, include blackholes, all the plastic ever produced, and radioactive plutonium. For each of these, as with climate change, one can see evidence of the hyperobject but never fully perceive the object itself.

Fossil fuel companies have taken advantage of the hyperobject nature of the climate problem to first deny and then minimize the risks with a well-funded, decades-long campaign to mislead and confuse Americans, and to falsely tarnish sensible proposals to address it as government overreach that would destroy jobs and undermine economic growth. Boomers have been the primary target of these campaigns, and an alarming number of us have fallen for these lies.

The impact of this disinformation can be seen not only on the far-right, among the media outlets like Breitbart and FOX News that deny climate change is an urgent problem. It's also evident in otherwise responsible mainstream media that have only recently begun to devote consistent, high-quality reporting to the issue. Even if you are an astute reader of *The Washington Post* or *The New York Times*, and your media diet includes centrist broadcasters like NPR and PBS, and even the nominally left-leaning MSNBC, it's quite possible you haven't been told what's in store if we fail to act.

For example, as I was writing this book, *The Washington Post*, my hometown newspaper, ran editorials supporting President Biden's plans to expand fossil fuel exploration in response to the energy crisis caused by Russia's invasion of Ukraine, rather than leading a World War II-style all-hands-on-deck approach to improving energy efficiency and expanding renewables and nuclear power. The editorial writers presumably know, yet somehow disregard, the scientific consensus that holding average global temperature to a level that humans can cope with requires that we build *no new fossil fuel infrastructure*. As terrible as it is, the war in Ukraine has not changed this imperative.

Let's Get Personal: Check Your Birth Year PPM

A quick and powerful way to understand how profoundly human actions have altered the atmosphere in our lifetimes is to check your birth year CO_2 ppm: the parts per million of carbon dioxide in the atmosphere the year that you were born.

You already know that CO_2 in the atmosphere traps heat, so more CO_2 means a hotter planet. You also know that burning fossil fuels, deforestation, agriculture, cement manufacturing and other activities generate CO_2 pollution. Before the industrial revolution, CO_2 hovered around 280 ppm for some 6,000 years, since the dawn of civilization. It then increased relatively gradually, reaching 310 ppm by the end of World War II in 1945. Since then it has zoomed off the charts, reaching 421 pm in 2022, and is still rising.

Scientists warned decades ago that the risks of severe and irreversible impacts increase substantially beyond 350 ppm. The pioneering climate advocacy organization 350.org takes its name from this number.

Checking the CO_2 ppm the year that I was born made these numbers personal, helping me to grasp how quickly the atmosphere has changed during my lifetime. When I was born in 1954, the CO_2 ppm was 318. The world first breached 350 ppm in 1986, when I was 32. By May 1992, when the Rio Earth Summit launched the United Nations Framework Convention on Climate Change (UNFCCC), I was 37 years old and the level was 359.99 ppm.

Knowing your birth year ppm puts you in this picture. The Nature Conservancy offers a simple tool: select your birth year and, presto, there's your answer. Go ahead, do it now on your phone: go to "nature.org," search for "ppm" and select your birth year.[5]

Some people, especially elders, have begun including their birth year ppm in their online bios. Inviting people in an online meeting to look up their birth year ppm and share it in the chat when they introduce themselves can be a good way to spread knowledge about this key indicator and get a sense of relative age without asking people to say how old they are.

UN Climate Reports Fall on Deaf Ears

Why are the basic facts about climate change—that the hour is late and the fate of humanity hangs in the balance—not more widely understood? One reason is that UN reports on climate change—the most comprehensive and authoritative summary of what to expect—often fail to attract more than passing attention in the U.S. media.

The Sixth Assessment of the Intergovernmental Panel on Climate Change (IPCC) published in a series of three reports in late 2021 and early 2022, with a capstone synthesis report planned for March 2023, should have been the exception. The first three reports confirmed what even casual news consumers had begun to fear: that the negative impacts of climate change are mounting much faster than scientists had predicted less than a decade ago. The reports showed that human-induced climate change is causing dangerous and widespread disruption in nature and affecting billions of people all over the world, with the people least able to cope being hardest hit.

What exactly is the IPCC? Although it is cited in the news each time it releases a report, its inner workings are little understood outside of the world of climate scientists and policy wonks. As we shall see, it has remarkable strengths and a possibly fatal flaw that has resulted in many interested people failing to grasp the urgency and severity of the problem.

Established in 1988 by the World Meteorological Organization (WMO) and the United Nations Environment Program (UNEP), and later endorsed by the UN General Assembly, the IPCC is headquartered in Geneva, Switzerland, and comprises 195 member states. Rather than conduct original research, it undertakes periodic reviews of all relevant published scientific literature. Hundreds of scientists and other experts from around the world review the data and compile key findings into Assessment Reports for policymakers and the public. With hundreds of scientists participating, it is arguably the largest scientific peer review process in the world.

Because the IPCC is governed by its member states, which elect a bureau of scientists to run it, it is a hybrid, part scientific body and part intergovernmental political organization. This structure was designed to ensure that the IPCC reports would be widely understood and endorsed by the global community. But it also has hindered the IPCC's ability to accurately assess the dangers and effectively sound the alarm.

Since 1990, the IPCC has issued a massive, multi-volume *Assessment Report* every five to six years. Recent assessments have included three working group reports—one on physical science; one on impacts and adaptation; and one on what must be done to slow temperature increases, what the scientists and policy experts call "mitigation." Each working group report has a *Summary for Policymakers* that can run up to a few dozen pages and must be approved, with line-by-line scrutiny, by all member states, and a *Technical Summary* that frequently runs close to 100 pages. At the end of each assessment, the three working group reports are summarized in a final *Synthesis Report*.

The length and complexity of these documents has given rise to a small army of IPCC report explainers—academics, journalists, and specialists in think tanks and other NGOs—who scrutinize each report and produce their own, much briefer, summaries, typically on impossibly short deadlines. With each successive assessment report, the science has become more certain and the warnings more dire.

The first working group report of the IPCC Sixth Assessment, on the physical science, was released in August 2021. It is the work of 234 scientists from 66 countries who consulted more than 14,000 scientific papers to produce a 3,949-page report, which was then approved by 195 governments. The *Summary for Policymakers* was drafted by the scientists and agreed to line-by-line by the 195 governments in the IPCC during the five days of negotiations leading up to the release.

The report warned that *massive and immediate cuts* in global emissions would be required to hold average global temperatures below a 1.5 °C (2.7 °F) or 2.0 °C (3.6 °F) increase from pre-industrial averages. These ceilings, endorsed as global goals in the 2015 Paris Agreement, were established in previous IPCC studies as *the maximum* that would be compatible with sustaining the natural systems that support civilization as we know it.

The second working group report of the IPCC Sixth Assessment, released in February 2022, was devoted to "impacts, adaptation and vulnerability" and was by far the most frightening yet. The UN Secretary General described it as "an atlas of human suffering and a damning indictment of failed climate leadership," adding, "with fact upon fact, this report reveals how people and the planet are getting clobbered by climate change. Nearly half of humanity is living in the danger zone—now."

Hans-Otto Portner, an IPCC co-chair, said: "The scientific evidence is unequivocal; climate change is a threat to human-wellbeing and the health of the planet. Any further delay in concerted global action will miss a brief and rapidly closing window to secure a livable future."

What should have been the final, urgent wake-up call to the world instead became a secondary story because, just four days before the report's release, Russia invaded Ukraine in a shocking war of aggression unlike anything that Europe had witnessed since the end of World War II. The resulting war included Russian rocket attacks on civilian shelters, the displacement of millions of Ukrainians, and the mobilization of

U.S.-led Western trade and financial sanctions against Russia. It also distracted attention from the climate threat, absorbing scarce international willpower that might otherwise have been marshaled to address the climate emergency.

While the war has the potential to hasten the end of the fossil fuel era—40% of the Russian government's budget comes from fossil fuel exports—fossil fuel companies and their enablers were quick to seize the opportunity to push for expansion of fossil fuel production. In the United States, Republicans blamed the Biden administration for rising gasoline prices, prompting even some moderate Democrats to propose cutting gasoline taxes—exactly the opposite of the policy urgently needed to speed the energy transition.

Unfortunately, the worsening impacts of climate change won't pause because the Russians invaded Ukraine. In mid-2022 the respected science journal Nature, one of many publications offering succinct summaries of the IPCC's working group report on climate impacts, highlighted the following conclusions:

- *Humanity will soon hit hard limits to its ability to adapt… For example, coastal communities can temporarily buffer themselves by restoring mangroves and wetlands but rising seas will eventually overwhelm such efforts.*
- *Climate change has already caused death and suffering… Smoke inhalation from fires has contributed to cardiovascular and respiratory problems, and increased rainfall and flooding has led to the spread of diseases such as cholera.*
- *If global temperatures rise by more than 1.5 °C above pre-industrial levels, some environmental changes could become irreversible.*

UN Reports Underestimate the Danger

As frightening as the IPCC predictions are, there's reason to believe that the climate impacts are more severe and are arriving sooner than the IPCC assessments predict. While climate-change deniers and so-called "skeptics" allege that climate scientists exaggerate their findings, historically the IPCC has *underestimated* the speed and severity of climate impacts.

There are two reasons for this. First, only research that has been accepted by a peer-reviewed scientific journal is included in the assessments. Because the lag between article submission and acceptance can often be one or more years, the assessments cannot include the latest science. This becomes increasingly problematic as the velocity of climate change accelerates.

Second, as we have seen, the Summary for Policy Makers, the part read by most journalists, think tank analysts, and climate advocates—often working on very tight deadlines—is subject to approval by IPCC member governments. The biggest and most powerful—the United States and the other big emitters—have plenty of reasons to downplay bad news. Close observers of the IPCC argue that, as a result, the summaries fail to convey just how bad things really are. The IPCC itself has implicitly acknowledged that its assessments have underestimated the rate of change, noting that impacts are more severe and arriving sooner than previously predicted.

Climate science pioneer James Hansen is among the scientists who believe that the IPCC is underestimating the threat. It was Hansen who first sounded the alarm in Congress, telling a Senate committee in June 1988 that "the greenhouse effect has been detected and is changing our climate now." Hansen was then the head of NASA's Goddard Institute for Space Studies and his testimony was the first warning about climate change to receive national news coverage. *The New York Times* ran the story at the top of the front page, with a graph showing a long-term rise in average global temperatures.

Now in his 80s and an adjunct professor at Columbia University's Earth Institute, Hansen continues to sound the alarm for all who will listen. In December 2021, soon after the UN climate conference in Glasgow, known as COP26, he published an article that was shockingly forthright:

> *Why is nobody telling young people the truth? "We preserved the chance at COP26 to keep global warming below 1.5°C." What bullshit! "Solar panels are now cheaper than fossil fuels, so all we are missing is political will." What horse manure! "If we would just agree to consume less, the climate problem could be solved." More nonsense!*

Hansen argues that the IPCC consistently underestimates the impacts already locked in. "It's certain that global warming will exceed 1.5°C and almost certain that it will exceed 2°C. That's what real data from the physical sciences tells us," he writes. Especially attention-grabbing is his contention that the IPCC has underestimated the impact of fresh water from melting polar ice, arguing that it will weaken ocean currents in ways that accelerate the melting of the Antarctic ice. If emissions continue rising at the current rate, Hansen warns, sea levels could rise "several meters within the lifetime of children born today." This would mean abandoning most of the world's coastal cities.

I've come to agree with Hansen and other independent climate scientists that the climate emergency is considerably worse than the IPCC, mainstream media and even the Big Green environmental NGOs who campaign for climate action let on. There's a seemingly sound reason for not telling it like it is: after all, people need hope to invest their scarce time and energy in averting catastrophe. Too much bad news can undermine the will to act.

But there's another reason as well. Because environmental organizations need money to operate, and because funders want to believe that their money will make a difference, big NGOs have a strong incentive to craft hopeful narratives about how this project or that program will make a difference.

But pulling our punches, failing to communicate the reality of the challenge, means that the proposed solutions are inevitably inadequate to the task. David Sprat, research director for the Melbourne-based Breakthrough National Centre for Climate Restoration, named this problem in a 2012 monograph, *Always Look on the Bright Side of Life: Brightsiding Climate Advocacy and Its Consequences*. "If you avoid including an honest assessment of climate science and impacts in your narrative, it's pretty difficult to give people a grasp of where the climate system is heading and what needs to be done," he wrote. "But that's how the big climate advocacy organizations have generally chosen to operate, and it represents a strategic failure to communicate."

Ouch. As a communications professional who devoted much of my career to informing people about climate change and global development, I have been complicit in this failure. It was only when young activists like Extinction Rebellion and Sunrise began insisting on the term "climate emergency" that the big green NGOs reluctantly adopted that terminology. For my role in that, I plead guilty.

Of course, things have only gotten worse since 2012. *Climate Dominoes*, the Breakthrough Center's 2022 report, calls the IPCC's Sixth Assessment "the most important assessment of humanity's future on Earth to date" but adds that it has a crucial blind spot: tipping points. "Major elements of Earth's climate system are now increasingly influenced by self-reinforcing warming processes—or positive feedbacks—due to climate change caused by human greenhouse gas emissions, mainly from burning fossil fuels," the report warns.

As with a row of dominoes standing on end, once the first domino falls it sets off a chain reaction, so that nothing can be done to stop the others from falling. "A 'tipping point' or critical threshold may be reached, such that a small change causes a larger, more critical change to be initiated, taking components of the Earth system from one state to a discreetly different state," the Breakthrough report explains. "Major tipping points are interrelated and may cascade, so that interactions between them lower the critical temperature thresholds at which each tipping point is passed." Such changes, the report notes dryly, "may be irreversible on relevant time frames, such as the span of a few human generations."

Alan Miller, Durwood Zaelke, and Stephen Andersen make a similar point in their book *Cut Super Pollutants Now,* in which they argue for rapid cuts in non-CO2 greenhouse gasses—such as black carbon (soot), methane, ground-level ozone, and hydrofluorocarbons (HFCs)—that have shorter lifespans in the atmosphere but much greater heating impact than atmospheric carbon.

> *Rapid warming over the near term threatens to accelerate self-reinforcing feedbacks in which the planet starts to warm itself in a Hothouse Earth scenario—vicious cycles, which could lead to uncontrollable warming as these feedback mechanisms become the dominant force regulating the climate system. These feedbacks would then set off a domino-like cascade that triggers tipping points in the Arctic and elsewhere, many of them irreversible and potentially catastrophic.*

For example, they explain, the area of the Arctic Ocean covered by summer sea ice shrunk by 40% from 1979 to 2011, raising the prospect that the Arctic could be without summer ice in as little as a decade. Ice reflects heat back into space, while blue ocean water absorbs it. The additional heat in the Arctic is setting off a second feedback loop, "accelerating the thawing of the permafrost in northern lands, releasing their ancient stores of carbon dioxide, methane, and nitrous oxide, causing still more warming."

Given these feedback loops and uncertainty about when specific tipping points occur, the Breakthrough report concludes, "the 1.5–2°C target range of the 2015 Paris Climate Agreement is demonstrably not a safe or appropriate goal for policymakers and advocates concerned about protecting the most climate vulnerable, and requires a major rethink about advocacy goals and what is possible and necessary to achieve them."

Boomers Don't Be Doomers

At this point, you may be thinking: given these terrifying facts and predictions, what's the point? But it's one thing to understand the urgency and severity of the threat and something quite different to give up and, even worse, to discourage others from acting. Boomer doomers who say it's too late to act elicit rage from younger people, especially climate activists, who know that they and future generations must live with the consequences of inadequate action today.

There is a fine line to walk between understanding the magnitude of the crisis and becoming overwhelmed—and overwhelming others. Doomerism—saying it is too late to do anything—can lead to the same conclusion as climate denialism: there is no reason to do anything.

Sarah Jaquette Ray, chair of environmental studies at California State Polytechnic University, Humboldt, and the author of *A Field Guide to Climate Anxiety*, is among the many climate communications specialists who warn that presenting the climate crisis as intractable can cause people to go numb and check out. To fight the sense of powerlessness, she said in a 2022 interview with *The New York Times*, she encourages people to see themselves as part of a collective groundswell of environmental groups working around the world, and to resist going down the rabbit hole of climate horror stories.[6]

While many impacts may be locked in or very close to a tipping point, organizing and fighting for a rapid end to the fossil fuel age and a transition to a fair, efficient, low-carbon economy can nonetheless delay or reduce these impacts. IPCC reports and other studies have shown a tremendous difference in impacts between an average temperature increase of 1.5 C and 2.0 C, with the additional half-a-degree resulting in much faster change and more severe impacts such as heat waves, droughts, flooding and extreme rainfall, sea level rise, and species loss.

Indeed, because scientists don't know the precise average temperatures at which planetary systems will cross various tipping points, *every tenth of a degree matters*. Slowing the rate of heating, even if the same average temperatures are eventually reached, also allows more time for societies and ecosystems to adapt, and for scientists to create new technologies—such as low-cost, highly effective direct air carbon capture and storage.

In the 2015 forward to the third edition of her 2003 book, *Hope in the Darkness: Untold Histories, Wild Possibilities*, Rebecca Solnit writes: "In the spaciousness of uncertainty is room to act. When you recognize uncertainty, you recognize that you may be able to influence the outcomes—you alone or you in concert with a few dozen or several million others. Hope is an embrace of the unknown and the unknowable, an alternative to the certainty of both optimists and pessimists."

Solnit makes clear that hope is different from denying the problem or pretending that things will work out for the best. Hope "is not the belief that everything was, is, or will be fine," she writes. "The evidence is all around us of tremendous suffering and tremendous destruction. The hope I'm interested in is about broad perspectives with specific possibilities, ones that invite or demand that we act."

Robin Wall Kimmerer, author of the best-selling *Braiding Sweetgrass: Indigenous Wisdom, Scientific Knowledge and the Teachings of Plants*, makes the case for joy as the antidote to doomerism. A scientist, decorated professor, and enrolled member of the Citizen Potawatomi Nation who was recently awarded a McArthur "genius" award, Kimmerer writes: "Even a wounded world is feeding us. Even a wounded world holds us, giving us moments of wonder and joy. I choose joy over despair. Not because I have my head in the sand, but because joy is what the earth gives me daily and I must return the gift."

Xiye Bastida, a Mexico-born youth climate activist and member of the Indigenous Otomi community, organized school strikes outside the United Nations headquarters in New York City in 2019, when she was just 17 years old. Bastida is among the many in her generation who are impatient with boomer doomers. "Our biggest problem is not denial. Our biggest problem is apathy, and that's what doomers are," she told a journalist at the time. "They write articles saying, 'We're all going to die,' so, 'OK, doomer.'"[7]

So, boomers, don't be doomers. Let's instead be climate boomers. While there is no denying that the world is already in very dangerous territory, and that this happened on our watch, there's still plenty of reason to fight. Knowing what to do requires understanding how we got here, and the forces that, even today, persist in blocking action. Given the high ideals that inspired us when we were young, what went wrong? We turn to that question in the next chapter.

Chapter 1: Are Boomers to Blame?

Key Takeaways

- Boomers shaped America's disastrous climate policy
- We all share responsibility, regardless of our role
- Climate change is more dangerous than you think
- UN reports fall on deaf ears *and* downplay the risks
- Even so, action now can still make a big difference
- Be a climate boomer, not a boomer doomer

What went wrong?

Some of us were tricked into denying the reality of climate change. Most of us knew but were complicit through inaction: we failed to demand that the United States lead a global transition to a low-carbon economy.

Chapter 2

What went wrong?

At the end of World War II, millions of Americans who had fought in Europe and the Pacific returned home to start families. These were the members of the Greatest Generation who had endured the Great Depression as they were coming of age and then saved the world from German and Japanese fascism. They had sacrificed greatly and those who were lucky enough to return home were ready to make babies.

"Suddenly pregnancy was patriotic," Landon Y. Jones writes in his 2008 book, *Great Expectations: America and the Baby Boom Generation*. "More than a million army wives had waited for President Truman to "bring the boys back home" and they and their husbands were celebrating with "a massive affirmation of childbearing."

More than 2.2 million couples recited marriage vows in 1946, twice as many as in any year before. The next year, more than 3.8 million babies arrived. The baby boom was under way. For nearly two decades from 1946 until 1964, an average of 4.2 million new babies were added to the population each year. As Jones writes: "They didn't know it but these ex-soldiers and their wives were making history at home in bed that would ultimately affect the country into the next century."

We boomers are the result of this post-World War II surge in births, prosperity and optimism. Before the war, U.S. fertility rates had been declining for decades. With the strong postwar economy, young couples felt confident that they could support several children. Plentiful jobs and GI Bill benefits, including college tuition assistance and subsidized mortgages in the burgeoning suburbs, made raising a family more attractive than ever before. With peace and prosperity, these young couples—

my young parents among them—had high hopes for their children.

"The boom babies were born to be the best and the brightest," Jones writes. "They were the first raised in the new suburbs, the first with new televisions, the first in the new high schools. They were twice as likely as their parents to go to college. They forced our economy to regear to feed, educate and house them. Their collective purchasing power made fads overnight and built entire industries... Blessed with the great expectations of affluence and education, the boom children were raised as a generation of idealism and hope."

In many fields, our generation lived up to those expectations. In *The Greater Generation: In Defense of the Baby Boom Legacy* (2006), Leonard Steinhorn argues that boomers transformed American culture to be broader-minded, more inclusive, and more tolerant, and motivated like no generation before us to care about the environment. A middle boomer, like me, Steinhorn packs his book with public polling data showing how progressive 1960s ideals went from outliers to mainstream. "Boomer youth culture became a conveyor belt that spread campus concerns to the rest of the generation, even to younger brothers and sisters too young to remember the Sixties," he writes.

Once the clamor from the Sixties had died down, we took the values from that decade and remade society "attitude by attitude, family by family, courtroom by courtroom, office by office, institution by institution," he explains. The result is nothing less than the transformation of American society:

> *Over the last four decades, the Baby Boom has created, reinvented, invigorated, or sustained most of the great citizen movements that have advanced American ideals and freedoms—the environment movement, the women's movement, the gay and lesbian movement, the human rights movement, the openness in government*

movement. In its wake, the Baby Boom generation has left not a single institution unchanged for the better, from the workplace to the university to the press to the military to the basic relationship between men and women."

Why, then, did we fail to act in response to the existential threat of runaway climate change? What went wrong? Part of the answer could be we were never as deeply committed to peace, justice and protection of the environment as the polling implied. More importantly, American corporations saw in the boomers a huge opportunity to sell stuff—if only we could be convinced that buying things was the path to happiness. The worst of these were the fossil fuel companies who waged a sustained, multi-million dollar propaganda campaign to undermine climate science. As we shall see, aspects of boomer culture made us easy targets for such lies.

This chapter tells the story of a generation of high promise that was too easily led astray, squandering the opportunity to prevent the climate emergency we now face. I hope that reflecting on what went wrong will help us to reconnect with the best values of our youthful selves, empowering us to do the right thing for our children, grandchildren, and the unborn generations who will come after we are gone.

Which Boomer Are You?

To be sure, we boomers are far from homogeneous. We are a big generation, and our views differ across all the dimensions that divide other Americans, including political affiliation, ethnicity and race, income, class, geography, religious background and even age.

Pew Research has shown that boomers have been about equally split over time between those who identify or lean Democratic and those who identify or lean Republican. And while people's political leanings tend to be stable over time, those who do change camps are more likely to shift from liberal to conservative than the other way around. This may help explain why a majority of us voted for Donald Trump both times he ran for president.

In addition, our coming-of-age experiences differed greatly by race. Roughly one out of ten of our generation are African Americans whose parents were often excluded from the GI Bill and other programs that underpinned the post-war boom. Historian Ira Katznelson has shown that the GI Bill "was deliberately designed to accommodate Jim Crow," by giving state and local officials the power to administer the funds.[8] Such discrimination was not confined to the South. Of 67,000 mortgages insured by the G.I. Bill in the New York and New Jersey suburbs, fewer than 100 were taken out by non-whites.

Climate change is sometimes seen as being of greater concern to upper-income, liberal white people. The polling suggests otherwise. Perhaps because low-income people are much more likely to suffer from fossil fuel pollution and climate impacts, Black Americans are much more likely than whites to say that they are concerned or alarmed about climate change. Hispanics have higher levels of concern and alarm than either group, with about two-thirds expressing such views. In contrast, conservative whites, opposed to what they see as government overreach, are often inclined to dismiss as unreal any problem that requires government policy to solve, making them susceptible to the climate change skepticism peddled by fossil fuel shills and amplified by conservative media.

Boomer experience and views also differ by age cohort. Many demographers distinguish between "leading-edge" boomers, born between 1946 and 1955, who came of age during the Vietnam War and Civil Rights eras, and the "late boomers," born between 1956 and 1964, whose formative memories include the Watergate scandal.

I'm a middle boomer, born in 1954, exactly the midpoint of our generation, too young to have participated directly in some of the seminal events, but old enough to have heard about them as they happened and sometimes to have watched them unfold on the TV news. For boomers younger than me, the watershed events of the late 1960s and early 1970s are things they heard about from their parents and older siblings, or even learned about in history classes.

1968: A Watershed Year

These differences notwithstanding, all of us came of age during a post-World War boom that combined sustained economic growth and dramatic, youth-led social upheaval. 1968 was a watershed year that seemed to unify an entire generation in a common sense of grievance against what we then called "The Establishment." The assassination of Martin Luther King Jr. in April sparked riots in more than 100 cities. Two months later, Robert F. Kennedy, younger brother of assassinated President John F. Kennedy, was shot as he campaigned in the California primary for the Democratic Party nomination. Although I was just 14 years old in 1968, like many older and middle boomers I remember where I was when I learned of each of these assassinations.

For many, Bobby Kennedy had embodied the hopes of our generation. He had championed Civil Rights and voter protection legislation as the attorney general in his brother's administration. He criticized the war in Vietnam, and during his campaign spoke eloquently on environmental issues, criticizing GDP as a deeply flawed measure of well-being. "It counts air pollution and cigarette advertising, and ambulances to clear our highways of carnage," he said. "It counts the destruction of our redwoods and the loss of our natural wonders in chaotic sprawl."

The tension between the young and the old came to a head at Democratic National Convention in Chicago in August 1968. Noisy boomer protests outside the convention hall were met with a violent police response. Inside, Hubert Humphrey, a liberal stalwart whose reputation had been sullied by serving as vice president to President Lyndon Johnson during the Vietnam War, won the nomination in a contentious vote seen as illegitimate by the protesters. In November, he narrowly lost to Richard Nixon, a Republican who ran on a promise to end the war.

The First Green Generation?

Two years later, in July 1970, our generation's prodigious political energy coalesced in the first Earth Day, a nationwide activity modeled on the "teach-ins" of the anti-Vietnam War movement, which attracted an estimated 20 million participants to events in universities, high schools, and even primary schools. Congress adjourned so members could speak at events across the country. Tens of thousands of people marched along New York's Fifth Avenue calling for environmental reforms.

Adam Rome's *The Genius of Earth Day: How a 1970 Teach-In Unexpectedly Made the First Green Generation* (2014) neatly captures the ethos and its legacy. Leading edge boomers helped to organize Earth Day events and younger boomers were swept along. "Thousands of organizers and participants decided to devote their lives to the environmental cause," Rome recounts. "Earth Day built a lasting eco-infrastructure: national and state lobbying organizations, environmental studies programs, environmental beats at newspapers, eco sections in bookstores, community ecology centers."

Before Earth Day, a series of disasters highlighted the urgent need for environmental safeguards. In November 1968, 78 miners were killed in a coal mine explosion in Farmington,

West Virginia. In January 1969, an oil well off the coast of Santa Barbara spewed 3 million gallons of crude that fouled beaches, killing thousands of birds and an unknown number of sea mammals. In June 1969, the polluted Cuyahoga River in Ohio burst into flames five stories high. The fire was a defining moment for the new environmental movement—and a nadir in U.S. environmental history—immortalized in "Burn On," a satirical song by Randy Newman that I and many boomers can sing to this day.

Recognition of the planet as a fragile, finite place was boosted by the first color photo of Earth from space, taken by U.S. astronaut William Anders aboard the Apollo 8, the first crewed voyage to orbit the Moon, in December 1968. With the gray moonscape in the foreground and the blue earth rising against the black of space, the image brought home to all who saw it—perhaps especially impressionable young boomers—the realization that humans have no other home. A LIFE book of 100 photos that changed the world called it "the most influential environmental photograph ever taken."

A Tide of Environmental Legislation

These events, and the upsurge in environmental activism sparked by the first Earth Day, propelled the issue to the center of the national agenda, opening the way for a tide of new environmental legislation unlike anything seen before or since.

In January 1970, President Nixon signed the National Environmental Policy Act (NEPA), which established a presidential Council on Environmental Quality (CEQ) and required for the first time that federal agencies proposing legislation or other actions that could significantly impact the environment provide an environmental impact statement, to be approved by the CEQ and made available for public comments.

In July the president signed an executive order creating the Environmental Protection Agency (EPA), which was ratified in committee hearings in the House and Senate. Within two years, the Congress had passed the Clean Water Act and the Endangered Species Act, both with large bi-partisan majorities.

The consensus for action to address environmental threats persisted into the mid-1980s. The discovery that chemicals used in air conditioners and aerosol sprays were destroying the ozone layer, which shields the earth from harmful ultraviolet rays, led to international action, with full U.S. participation and support. Signed in 1987, the Montreal Protocol was the first treaty in the history of the United Nations to achieve universal ratification.

Crucially, U.S. chemical companies that manufactured the ozone-destroying chemicals—hydrofluorocarbons—had a technological lead in creating and supplying less-damaging alternatives, and therefore supported the treaty. The situation stands in stark contrast to the stance of America's fossil fuel companies, who see renewable energy as competition and have blocked the transition to a low-carbon economy at every possible turn.

As late as the early 1990s, however, climate change was not yet a partisan issue. It was a Republican president, George H.W. Bush (Bush the Elder), who signed the United Nations Framework Convention on Climate Change (UNFCCC) at the Rio Earth Summit in 1992, a treaty that was subsequently ratified by a bipartisan vote in the Senate. The Convention committed signatories to combat "dangerous human interference with the climate system" by stabilizing greenhouse gas concentrations in the atmosphere.

It is the UNFCCC, which has a secretariat in Bonn, Germany, that convenes the annual UN climate gatherings known as the "Conferences of the Parties" or COPs. The UNFCCC managed the negotiations that led to the 1996 Kyoto Protocol and the 2016 Paris Climate Agreement, which are technically implementing measures of the UNFCCC. As we have seen, however, neither of these would be ratified or otherwise endorsed by the Senate.

Fossil Fuel Companies React with Lies

Fossil fuel companies have long known that their products threatened climate stability. They hid this knowledge, instead launching misinformation campaigns designed to block efforts to regulate greenhouse gases and otherwise restrain their industry.

The first big campaign was led by the misleadingly named Global Climate Coalition (GCC), a front group for fossil fuel companies and allied business interests that worked to block the regulation of greenhouse gases. Founded in 1988, it lobbied government officials, criticized international climate organizations, highlighted the uncertainty in climate models, and verbally attacked scientists and environmentalists. It minimized the risks of climate change and falsely asserted that there were trade-offs between economic growth and emissions reductions.[9]

Although bad publicity compelled companies to withdraw from the GCC—Green Party delegates to the European Parliament proposed naming hurricanes after member companies—other groups were quick to take its place.

Several excellent books have told how fossil fuel companies and trade associations such as the American Chamber of Commerce, the National Association of Manufacturers, and the American Petroleum Institute—abetted by fake think tanks like the Heritage Foundation and the Heartland Institute—systematically undermined the consensus for climate action.

Merchants of Doubt (2011) by Naomi Oreskes and Erik Conway shows how companies, putting their short-term interests first, used an ideology of free-market fundamentalism and a too-compliant media to confuse Americans about a wide range of threats to our health and well-being, from DDT and tobacco to acid rain and, most consequently for the future of civilization, climate change.

We boomers, who by this time were the majority of American voters and held most of the seats in Congress, were the main target of these efforts. Some of us fell for these lies, rejecting climate science because we had been persuaded that transitioning from fossil fuels would threaten our jobs and lifestyles. Others became complicit through inaction: we knew that climate change was a real threat but, busy working and raising our kids, we failed to speak loudly enough and organize well enough to overcome these powerful vested interests.

Such was the political landscape at the end of 1996 when Vice President Al Gore, a leading-edge boomer, began pushing for Senate ratification of the Kyoto Protocol, which he had helped to negotiate. It was not to be. The next year, a Senate elected by boomers with a majority of boomer members, voted unanimously for a resolution that implicitly rejected the agreement, stating that any treaty that imposed mandatory emissions targets on industrialized countries without also doing the same for developing countries "would result in serious harm to the economy of the United States."

Developing countries had rejected targets for themselves, pointing out that the industrialized countries, especially the United States, had become prosperous burning fossil fuels. The big historical emitters, they argued, were responsible for the vast majority of the CO_2 accumulated in the atmosphere and were therefore morally obliged to cut emissions first.

The distinction between *accumulated* emissions and *current* emissions, fundamental to climate justice, is one that U.S. officials, politicians and news outlets consistently underplay, even today. Although China became the world's largest *current* emitter in 2006, the United States, which industrialized much earlier, is responsible for twice as much of the accumulated emissions now heating the planet. With a mere 4.25% of the world's population, the United States accounts for 25% of accumulated emissions. Moreover, current U.S. emissions of 14 tons per person are twice those of the average Chinese.

Unaware of America's responsibility for the problem and therefore our obligation to lead, boomers stood by while the fossil fuel companies and their lobbyists torpedoed the Kyoto Protocol. Subjected to a constant bombardment of sophisticated disinformation, we had gone from being youthful idealists organizing Earth Day teach-ins to being a solid middle-aged, middle-class and increasingly self-absorbed majority that failed to rally in support of international action that could have prevented the crisis.

The Greening of America—Not So Green After All

Why did we boomers, "the first green generation," fall for these lies?

I believe the answer lies in boomer values celebrated in a once hugely influential book, *The Greening of America*, by Charles Reich. Published in 1970, the same year as the first Earth Day, it sold millions of copies, and spent weeks on *The New York Times* best seller list, including three weeks at No. 1. I remember reading it as a teenager riding a cross-country Greyhound bus, eagerly underlining passages that seemed vitally important to me and thinking "yes, this is me, these are my people, this is my generation."

It is difficult now to overstate how influential this book became in shaping the understanding of our generation. The original version, a 25,000-word essay in *The New Yorker*, elicited more letters from readers than the magazine had ever received before. Copies were scarce and it eventually went through 20 printings. *The New York Times* alone published about a dozen articles on the book.

David Skinner, editor of *Humanities* magazine, nicely summarized the arguments and lasting influence of *The Greening of America* in a 2005 article marking the 35th anniversary of its publication:

> *War and poverty, uncontrolled technology and the destruction of the environment, the Corporate State and bureaucracy, the artificiality of work and culture, the absence of community—all had conspired to produce the most "devastating" impoverishment of all, the "loss of self, or death in life."*
>
> *Yet there was hope, for the crisis was calling forth its own antidote: a movement to reclaim "a higher reason, a more human community, and a new and liberated individual." That movement—which Reich predicted would eventually grow to include all Americans—was none other than the youth culture of the 1960s.*

Reich (the "ch" is pronounced "sh") was a 40-year-old Yale Law School professor and former Supreme Court clerk when he headed to San Francisco in 1968 for what became known as the Summer of Love. There he imbued the counterculture values of the young boomers who had flocked to the city, placing them within a grand theoretical construct. The young, he wrote, had created a "new level of consciousness" which, unshackled by the stifling moral constraints of the 1950s, focused instead on spiritual fulfillment.

"The extraordinary thing about this new consciousness," Reich wrote, "is that it has emerged from the machine-made environment of the corporate state, like flowers pushing up through a concrete pavement. For those who thought the world was irretrievably encased in metal and plastic and sterile stone, it seems a veritable greening of America."

"Green" was not yet synonymous with the things that are good for the environment, much less with climate action. Instead, the book took aim at conformity. It celebrated the choice of an alternative lifestyle and the talismans of the new generation: sex, drugs, rock-and-roll, frisbee, organic peanut butter and especially blue jeans. Blue jeans, inexpensive to buy and maintain, were heralded as a symbol of the liberating spirit of the new generation that would lead America into an anti-materialist, community-minded, freedom-loving era of love, peace and hope.

Except it didn't quite turn out that way. The arrival of designer blue jeans as expensive status symbols is sometimes held up as proof of boomer hypocrisy. To me, it's also a powerful symbol of corporations' success in commodifying and co-opting an emerging youth-led counterculture.

"For better or worse, designer jeans encouraged the consumer to look at premium products not just as an extravagance but as an investment," culture critic James Sullivan wrote in *Jeans: A Cultural History of an American Icon.* "Every accoutrement of the American lifestyle has been branded, and it started with the brands on our behinds."

Boomers Fall Prey to the Century of the Self

So, did Reich get it wrong? I think only partly. His description of boomer values was largely spot-on: side-by-side with the commitment to social and environmental justice was a focus on peak experiences, self-actualization, and pleasure. The boomer slogans "Be here now!" and "If it feels good, do it!" didn't lend themselves to caring about the future or taking up the mantle of global leadership.

Similarly, our distrust of what we called "The Establishment" and the slogan "Don't trust anyone over 30!" left some of us inclined to doubt experts and science, making us vulnerable to baseless conspiracy theories. Trump's boomer supporters—with their climate change denialism, gullibility to false claims that the 2020 election was stolen and baseless objections to public health measures to counter the COVID-19 pandemic—

are the dark side our generation's preference for intuition over rationality.

Today our hedonism is nowhere more apparent than at The Villages, the sprawling retirement community in central Florida, where booze, golf, sex, and rock-and-roll are the dominant pastimes. According to a March 2022 story in *The New York Times*, among the 84,000 residents, Republicans outnumber Democrats by more than two-to-one.[10] The fact that much of the state's low-lying coast will be underwater within the lifetime of their grandchildren evidently causes these Florida residents little concern.

How did boomer attitudes travel so far, from the first Earth Day to life in The Villages? To use the language of the 1960s, we were co-opted by the great American marketing machine. The fossil fuel lobby's disinformation campaigns were part of a larger effort to persuade us that self-fulfillment lay in material goods, not experiences and community.

The mechanics of this deceit is vividly described in *The Century of the Self*, a 2002 British television documentary by Adam Curtis that tells the story of Edward Bernays, a German-born American who pioneered what came to be known as public relations. A nephew of Sigmund and Anna Freud, Bernays drew upon their insights into human nature to harness our generation's desire for self-actualization to shore up capitalism.

As Curtis tells it, corporate leaders came out of World War II worried that the new system of mass production would produce more goods than people wanted to buy. The documentary quotes a Lehman Brothers executive who was frank about the solution. "We must shift America from a needs to a desires culture," the executive wrote. "People must be trained to desire, to want new things even before the old had been entirely consumed. We must shape a new mentality in America. Man's desires must overshadow his needs."

Most of us, whether we went on to become Trump Republicans or Clinton/Obama/Biden Democrats, fell for this, hook, line and sinker. We were put off by what we viewed as artificiality in work and culture, and we longed for community. But for many of us, seduced by the constant bombardment of marketing messages, it was but a small step from self-actualization through long hair, blue jeans, sex, drugs, and rock-and-roll, to self-actualization through giant houses, giant SUVs, exotic vacations, and burgeoning tax-free retirement savings accounts.

Reconnecting with Our Youthful Values

Can enough of us reconnect with the better aspects our youthful values—our concern for social justice and protection of the natural world—to make a difference in the fight to avert climate catastrophe?

I agree with Steinhorn's assertion that our generation's core values are real, correct and now widely held. The millions of us who protested the Vietnam War really did want peace, and we still do. The millions of us who participated in the first Earth Day, and many of the Earth Day celebrations since, really do want to protect the natural world and pass on to future generations a livable planet.

The challenge is to reconnect with idealism of our youth and use it to turbo-charge our activism. We must recognize that we were deceived into thinking that self-actualization lay in what we buy and own, not our relationship with others. We must reignite our willingness to challenge authority and deploy our power against those who would destroy the planet in their pursuit of profit. We must also embrace our role as elders and ancestors, responsible for passing along a livable planet to those who come after us.

Fortunately, there are millions of boomers who are prepared to do this. The next chapter assesses our power.

Chapter 2: What Went Wrong?

Key Takeaways

- We began with high ideals, the first Green Generation
- Earth Day sparked sweeping environmental laws
- Fossil fuel companies reacted with lies
- Advertisers encouraged our pursuit of pleasure
- Now is the time to reconnect with the idealism of our youth

Chapter 3

Can boomers avert catastrophe?

We are still the world's wealthiest and most powerful generation—and our power is growing as we stop working and have more time to spend on politics. If enough of us mobilize now we can help the world avoid the worst impacts of climate change.

Chapter 3

Can boomers avert catastrophe?

We have seen that we boomers failed to live up to our youthful ideals during the crucial years when climate change went from a distant threat to a planetary emergency. The good news is that it's not too late for our generation to play a central role in averting an eons-long catastrophe. Indeed, we are uniquely placed to make a difference. Although our generation's power and the global power of the United States have both declined since their peaks, we continue to be the most powerful generation in the most powerful nation on earth.

Of course, there are caveats. As we shall see, not all of us are willing or able to join in the growing movement for climate action. Nonetheless, tens of millions of us are coming to understand the threat and have both the means and the inclination to act.

This chapter concludes Part I with an assessment of boomer power. Part II turns to what we can do, starting with simple steps that we can undertake on our own (Chapter 4), by joining with others (Chapter 5), as members of faith-based communities (Chapter 6), by participating in non-violent civil disobedience (Chapter 7), and by supporting a radical flank (Chapter 8). I conclude with a short summary and three tips for new boomer climate activists (Chapter 9).

I wrote this book because I believe that our generation's broad-based active participation in pushing for climate action is crucial to averting a planetary catastrophe.

One reason is a shortage of time. We have seen that humanity has less than a decade to drastically reduce emissions to avoid a cascade of feedback loops that would threaten civilization as we know it. Humanity can only dodge the climate catastrophe bullet if those of us alive today—including many of our generation—demand that our leaders act. We simply can't leave it to future generations to fix the problem. The compounding impacts of cumulative emissions mean that by the time they come along, it will be too late.

The other reason is generational power: Even as we reluctantly relinquish the stage to the generations after us, we remain, for now, the most powerful generation in the most powerful nation on Earth.

Which Generation Holds the Most Power?

By multiple measures, we are the wealthiest and most influential generation on the planet. In 2021, Visual Capitalist, an online publisher of data visualizations, crunched the numbers to create a Generational Power Index (GPI) and concluded that boomers held 39% of power in the United States, followed by Gen X, with 30% and millennials with 15%. The generation ahead of us, the so-called Silent Generation, was hanging on with 13% of power.[11]

The GPI calculated the generational shares of three types of power—economic, political, and cultural—and concluded that we hold the most economic and political power, while ranking second only to Gen X in cultural power. We have retained our top spot even as the millennials surpassed us as the country's largest living adult generation. By 2021, millennials (born 1981–1996) numbered 72 million, two million more than us boomers. Gen X (born 1965–1980), numbered 66 million, while Gen Z (born 1997–2012), the youngest named U.S. generation, clocked in at 69 million.

Since the GPI was compiled, turnover in the Supreme Court, the retirement of some prominent boomer politicians, and arrival on the scene of younger members of Congress elected in the 2022 midterms have somewhat eroded boomer political power. But experts believe that our generation will continue to have power beyond what our numbers warrant at least until the end of this decade.

In *Generation Gap: Why the Baby Boomers Still Dominate American Politics and Culture* (2022), political scientist Kevin Munger argues that "far from going away, the power of the boomers will increase over the next five years as more of them retire and spend more time and energy participating in electoral politics and consuming culture and media. At the earliest, boomer power will peak sometime in the late 2020s." We will retain power, he writes, because of "an unprecedented concentration of raw demography, wealth, cultural relevance, and accumulated historical experience in a single generation at the top of the age distribution."

Munger calls this phenomena "boomer ballast." As with the ballast in a ship, it makes for stability but also makes it hard to change course. "Our ship of state thus has more ballast than ever before, rendering us unusually stable and slow to adapt," he writes. While younger generations are concerned about student debt, housing prices and climate change, boomers tend to be more worried about pensions, Social Security, and Medicare. Not surprisingly, "the dominant perspective of many young activists is that they wish boomers would just go away," he writes.

With great power comes great responsibility. Since we aren't going away, it's all the more crucial that those of us who recognize the climate emergency and feel a moral obligation to act understand our generation's power and deploy it as well as we can to help address this existential threat.

Boomers Have the Most Economic Power

To calculate economic power, the GPI weighs four indicators: net worth, median earnings, billionaire wealth, and business leaders. Boomers come out on top for all four measures except median earnings, where we are tied with the Silent Generation, at 22% of the pie. When the index was compiled, more than half of business leaders were boomers and our generation held $18.5 trillion in equities and mutual funds, more than twice the holdings of the next richest generation, the Gen Xers. Shockingly, we had more economic power overall than the three younger generations combined.

One reason we come out on top in net worth: we own the lion's share of the nation's real estate. Our real estate holdings surpassed those of Silent Generation in 2001, peaked ten years later at nearly half, and then declined only slightly, to 44% in 2021. Our preference for aging in place suggests that our share of the real estate pie probably hasn't shrunk much since. Our vast real estate holdings imply responsibility and opportunity: with residential buildings accounting for nearly a third of U.S. emissions, those of us who stay in our homes can help to reduce emissions by improving insulation and installing rooftop solar, as we discuss in the next chapter.

Boomers Wield Immense Political Power

When it comes to political power, boomers "are still the clear leaders in the political arena, but power dynamics are beginning to shift," according to the GPI. The index tallies five variables to calculate political power: number of eligible voters in each generation, political spending, and federal, state and local positions in government. In 2021 we came out on top in all categories except local positions, where the Gen X held 46% of the positions compared to our 39%.

The 2022 midterms resulted in a Congress that for the first time did not have a boomer majority. Nonetheless, boomers retained a big plurality: 48.5% of senators, representatives and delegates in the 118th Congress are boomers, followed by 35.6% Gen Xers, and 10.2% millennials. Rep. Maxwell Frost, a 25-year-old Democrat from Florida became the first Gen Z member of Congress. The median age in the House dropped a year to 57.9

years, while the median age in the Senate—where 66 of the 100 seats are held by boomers—ticked up nearly a year, to 65.3 years.[12]

While a shift has clearly begun, our generation is likely to remain disproportionately powerful for some years to come. Munger points out that what counts is not the number of eligible voters but turnout. We are much more likely than younger generations to cast ballots. For example, although turnout among young adults surged in the 2020 elections, boomers still cast more votes than any other generation. Pew Research estimated that 36% of those who voted that year were boomers, followed by Gen Xers at 25% and millennials at 22%. Munger also notes that boomers have more free time and more money to devote to politics than younger people, giving us disproportionate clout, even as the torch passes to younger politicians.

Boomers Cultural Power Now Second to Generation X

The GPI measures cultural power with a grab bag of eight types of indicators—digital platforms, films & TV, books, sports, art, celebrities, music & radio, and press & news media. It finds that boomers, who long dominated American culture, now have 25% of generational power, far behind Gen X with 36%. We come out on top in two old-school categories: books (authors of best-selling books) and art (top artists by auction revenue and influence). The GPI predicts that boomers and Gen X will soon be overtaken by the millennials, who are already nipping at our heels with 24% of cultural power, because of their much greater digital media savvy. When that happens, our once vast influence over the nation's culture will drop to third place.

Munger takes a more sophisticated approach to cultural power, showing that different generations consume media, especially news, through very different platforms. Cable channels and a handful of programs on broadcast television still dominate the news media landscape for boomers and Gen

Xers. He cites a Pew Research Center survey showing that more than 80% of people over 65 often get their news from television, compared to only 16% for adults under 30.

These alternative media ecosystems will increasingly diverge, he writes. "Demographic trends suggest that the relevance of Fox News and MSNBC will increase in the short run" as boomers retire and spend even more time watching television. But young people are neither watching these outlets nor waiting around for their chance to become a cable talking head. Instead, they are building a new ecosystem using YouTube and various social media platforms. So even as boomer dominance of cable and other legacy media approaches peak, "the alternative media ecosystem will grow, largely unseen by the legacy media and those who fixate on it."

The GPI reaches a conclusion that few generational scholars dispute: "Boomers wield the most power of any generation—for now." With time, of course, our power will surely diminish. "As wealth is passed on to children, younger generations move into political positions, and digital media continues to gain ground, the power dynamic is constantly shifting," the GPI notes. "Will it be Gen X or millennials who eventually move into the top spot?"

For those who study generational dynamics, that is indeed an interesting question. For us boomers—and for the future of civilization—a more important question is this: how will we wield our vast generational power during the remaining ten years that we still have it?

Properly Mobilized, Half of Boomer Power Is Still a Lot of Power

Of course, we boomers are diverse in our views, not only on climate action but on a host of hot-button issues. The Red-Blue partisan divide is perhaps greater among boomers than any other generation. And contrary to the progressive reputation we earned in our youth, in recent elections most of us backed the Republican Party, which has consistently blocked climate action.

In 2016, boomer voters favored Trump over Clinton by 50% to 46%. The margin narrowed slightly in 2020; yet we still went for Trump by a 3-point margin. Shameful, to be sure, given what the country had been through in the previous four years, but less reprehensible than our Silent Generation elders, who favored Trump by a shocking 19 points in 2016 and 16 in 2020.

Regardless of our political views, a significant number of us struggle just to get by. For older people worried about paying for rent, medicine and groceries, finding time and energy for social causes, even causes as pressing as the climate emergency, can be a challenge. One recent study found that 40% of boomers had no retirement savings; of those who do have retirement savings, one-out-of-four had less than $10,000.

Some of this is due to timing. The oldest boomers began retiring in 2011, just two years after the 2008–09 financial crisis halved the value of the stock market. Although the markets surpassed pre-crash highs in 2013, retirees who panicked and sold assets have struggled to recover. Of course, many poor and working-class boomers never had much retirement savings to lose in the first place, due to rising inequality and stagnant wages. For boomers of color, systemic racism has eaten away at earnings, savings, home values and overall wealth throughout their working lives, leaving some to face severe financial insecurity in old age.

These caveats notwithstanding, plenty of boomers are doing just fine. Indeed, the winners-win, losers-lose pandemic economy has resulted in many in our generation entering retirement with more savings than they expected.

"For better-off Americans, the pandemic economy created some of the strongest incentives to retire in modern history, with generous federal stimulus, incredible market gains, skyrocketing home values and health concerns drawing many Americans into early retirement," *The Washington Post* reported in late 2021. Families that entered the pandemic with a nest egg of about $700,000 could easily have become millionaires, even without counting the value of their homes, the article noted. While the 2022 market decline has since trimmed those gains, millions of boomers are already enjoying or can look forward to a comfortable retirement.

What about attitudes? That more than half of our generation voted twice for Trump suggests that many of us are really gullible—or chose for ideological reasons to believe lies. Yet polling shows that an overwhelming majority of us understand that climate change is real and urgent, and support government action to address it. Although the proportion of us holding these views is lower than for younger generations, there are still millions of us who understand the issue and, depending on the circumstances, could be persuaded to act.

For example, according to a 2021 Pew Research study, 57% of Americans who are "boomers and older" say that climate should be a top priority for assuring a sustainable planet for future generations. While this falls short of the 71% of millennials professing this view (as well as the 67% of Gen Z and 63% of Gen X), that still means that there is a solid majority, almost 40 million boomers, who hold this view. Fully 29% of boomers and older said that addressing climate change is their "top personal concern." And while that's 8 points below Gen Z, it still means that one out of three us born before 1965 consider climate their top concern.[13]

Climate Change in the American Mind, a series of nationally representative surveys conducted jointly by experts and George Mason University and Yale University, has found similar trends. In a 2019 study, 63% of boomers said that global warming was personally important to them, lower than the 73% of millennials, but still two out of three of us. Fully a third of boomers said that they were willing to contact government officials about the issue or donate money to an organization working to address the problem.[14]

Moreover, our generation's attitudes are probably shifting in favor of climate action along with the rest of the country as the impacts become ever more frightening. *Climate Change in the American Mind* groups Americans' views into six groups: dismissive, doubtful, disengaged, cautious, concerned, and alarmed. In late 2021, the percentage of "alarmed" suddenly surged to 33%, the largest share by far, followed by "concerned" with 25%—and that was *before* the punishing summer of 2022. While a breakdown of attitudes by generation is not available, given the size of the surge is seems likely that a growing number of boomers are among those waking up to the threat.

In summary, we retain a lot of power, being the country's most powerful generation in the realms of economics and politics, even as our cultural power slips behind Gen X and may soon be pushed into third place by those digital natives, the millennials. And while we are far from a consensus on climate change—most of us voted for a lying climate-change denier in 2016 and 2020—about two-thirds of us nonetheless tell pollsters we regard climate change as personally important.

In addition, we may be more ready to act on these views than is generally recognized. In a September 2021 poll by the *Climate Change in the American Mind* program found that about one out of five boomers would definitely or probably support an organization engaged in non-violent civil disobedience to push for climate action, the same proportion for Gen X (both were more than ten points behind millennials and younger generations, 35% of whom would definitely or probably support such organizations).

Fully 10% of boomers said that they would definitely or probably participate in non-violent civil disobedience (defined in the survey as "sit-ins, blockades or trespassing") against government or corporate activities that make global warming worse, if a person they like and respect asked them to do so. Ten percent of 70 million boomers means that some 7 million boomers are open to such a course of action, while fully 2 million said they would "definitely" do so.

Perhaps surprisingly, given that Black Americans have much worse experiences with police than white Americans, Black people of all generations were about twice as likely as white people to say that they would definitely or probably participate in non-violent civil disobedience to oppose actions that would make climate change worse (22% compared to 11%).[15] This may be because climate change disproportionately hurts poor people and people of color, as well as a greater appreciation of the role of non-violent civil disobedience in the Civil Rights movement. Whatever the reasons, given that Black families on average have much less wealth and income than white families, it suggests Americans need not be well-off to be willing to take personal risks to push for climate action.

Of course, nowhere near this number of people, of any race or generation, have consistently mobilized for climate action, much less risked arrest. The gap between saying you are willing to act and actually doing something is larger for our generation than for any other. According to Pew, although 27% of boomers say that addressing climate change is their top personal concern, only 21% have personally taken action to help address climate change in the past year. That 6-point gap was larger than the 5-point gaps for millennials and Gen Z, and 50% bigger than the 4-point gap for Gen X, suggesting a large untapped potential for boomers to become climate activists.[16]

What's in It for Me?

We have seen that our generation has a moral responsibility to step up our climate activism, and that millions of us have the means and the inclination to do so. And yet... although few of us would dare say this out loud, many of us nonetheless wonder privately: "What's in it for me?" This is hardly surprising. The focus on "me" is human nature—and may be especially true of our generation. Gonzo journalist Tom Wolfe famously dubbed the 1970s the "Me Decade" for a preoccupation with self-fulfillment that verged on narcissism, leading other social commentators to dub us boomers the "Me Generation."

For many of us, this private search for meaning continues and, indeed, takes on renewed urgency as we leave behind the working and child-rearing stages of our lives. Ironically, caring more about others—specifically people impacted by climate change who may be in our own communities, or in far-away lands, or even yet unborn—may turn out to be the answer.

In 2017, Harry R. Moody, a renowned expert on aging and the author of *The Five Stages of the Soul* (1998), published an article in *Journal of the American Society on Aging* titled "Baby Boomers: From Great Expectations to a Crisis of Meaning." "If we live long enough, some disillusionment is inevitable, and aging brings its share," he wrote. "But baby boomers grew up with great expectations for the future—for themselves and the wider society... Aging baby boomers will have to find a new story for themselves as they search for meaning in a world different from what they expected."[17]

Moody is a lucid and well-read writer who quotes liberally from those he has read to weave a tapestry that presents a picture all his own. He quotes Danish author Isak Dinesen: "all the sorrows of life are bearable if only we can convert them into a story." Moody adds that we boomers, "individually and collectively, will face the challenge of how to convert their experience into a story that makes sense."

How to craft this story? In answer, Moody quotes Viktor Frankl, the famed Austrian neurologist, psychiatrist, philosopher and Holocaust survivor, who argued that the search for life's meaning is the central human motivational force. Frankl, Moody tells us, answered questions about the meaning of life by drawing on lessons he learned in the Nazi death camps. Frankl said he discovered that "it did not really matter what we expected from life, but what life expected from us. We needed to stop asking about the meaning of life and instead think of ourselves as those who were being questioned by life—daily and hourly. Our question must consist not in talk and meditation, but rather in right action and in right conduct."

Aging baby boomers, Moody concludes, writing now in his own voice: "will have to find a new story, one that not only makes sorrow bearable but also converts our experience into hope for future generations."

As we face a narrowing window for action to avert a planetary catastrophe that will last for countless generations, this must surely mean doing all we can to avert that catastrophe. The remainder of this book tells how.

Chapter 3: Can Boomers Avert Catastrophe?

Key Takeaways

- Boomers are still America's most powerful generation
- We have economic, political and cultural power
- Not all boomers get it: half of us voted for Trump—twice!
- But the other half of boomer power is still a lot of power
- Acting on climate can help bring meaning to your life

Part 2

What we can do about it

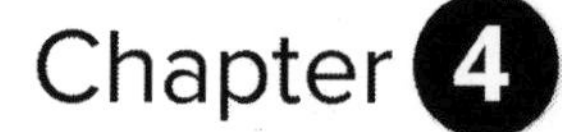

Every little bit helps, right?

Here are six individual actions that can make a difference. But don't let personal action distract you from helping to demand rapid, systemic change.

Chapter 4

Every little bit helps, right?

Does it make sense for boomers to take personal action to address the climate emergency? For many of us, this seems the logical place to start. After all, we must start somewhere and these actions are the only things we can fully control ourselves.

There are scores of books and thousands of articles about small, individual actions you can take to slow climate change. Yet we know that avoiding catastrophe requires rapid, systemic change, a transformation of societies and economies unlike anything the world has ever seen. And this must be achieved in the face of determined opposition from powerful, wealthy people who benefit from the status quo, as well as ordinary people who have been persuaded that climate action would threaten their jobs and lifestyles. In the face of such opposition, can personal and household actions make a difference?

For boomers there's an additional factor: most of our emissions history is behind us. Even if you died tomorrow, ceasing your emissions entirely, the greenhouse gasses you caused to be emitted during your lifetime will go on heating the planet for centuries to come.

Then there's the risk of a cognitive bias called moral licensing, in which people take small actions to fix a big problem, decide that they have done their bit and do no more. I think of this as the tote bag bias: a person drives to the store in a SUV, stocks up on groceries including plenty of meat, then feels virtuous because they take it all home in reusable totes.

Corporations encourage tote bag bias to divert attention from the need for effective policies and regulations so they can continue their business as usual. Many boomers remember the 1971 TV ad in which an Italian American actor dressed as a

Native American paddles a canoe along a polluted river. A bag of trash thrown from a passing car lands at his feet. A tear rolls down his cheek. The kicker, intoned by an off-screen narrator: "People start pollution. People can stop it."

The ad was created for Keep America Beautiful, the public relations arm of beverage and packaging companies that were pumping out billions of plastic bottles each year while blocking a law that would have required a 5-cent deposit on beverage containers, as well as other environmental legislation. The goal of the campaign, wrote Finis Dunaway, a professor of environmental history, was to "conflate litter with pollution, making the problems seem indistinguishable from one another… [while] making viewers feel guilty for their own individual actions."[18]

Less well known, even among people who passionately support climate action, is that one of the world's biggest oil companies created the concept of "personal carbon footprint" for much the same reasons. Mark Kaufmann, the science editor at *Mashable*, relates how 20 years ago British Petroleum hired the public relations firm Ogilvy & Mather to persuade people that climate change is our fault, not the fault of the giant oil companies that extract, process, and sell the stuff, propped up by billions of dollars in tax-payer subsidies, while they block policies that would speed the transition to lower carbon alternatives.[19]

BP unveiled its "carbon footprint calculator" in 2004 so that people could assess "how their normal daily life… is largely responsible for heating the globe," Kauffman writes. "A decade and a half later, 'carbon footprint' is everywhere. The U.S. Environmental Protection Agency has a carbon calculator. *The New York Times* has a guide on 'How to Reduce Your Carbon Footprint.' *Mashable* published a story in 2019 entitled 'How to shrink your carbon footprint when you travel.' Outdoorsy brands love the term."

We must be cautious to avoid falling for such propaganda, allowing individual action to distract us from the urgent need for systemic change. At the same time, some individual actions can indeed help to build momentum. The trick is to choose these actions wisely and not stop there, to avoid falling for tote-bag bias.

Environmental economist Gernot Wagner knows the folly of relying on personal action as a solution better than most: he has written an excellent book arguing that solving environmental problems requires smart policy rather than personal action.[20] Yet Wagner and his wife recently spent a lot of money to reduce their climate footprint, refitting their pre-WWII New York City co-op to save energy and cut emissions.[21] The retrofits, which cost $100,000, cut their electric bill, from about $400 a month to $100 a month. Of course, as he notes, "it will take decades to recoup our initial outlay." Why bother?

When I asked Wagner this, he pointed me to an article he wrote for *The Economist*. For individual actions to be effective, he writes, "it is essential that they generate momentum." Bicycling, he writes, is one example. People who ride bikes urge their local governments to build more bike lanes, which makes other people feel safer riding bikes, which leads to more bike riders, which leads to more and better bike infrastructure, creating a virtuous loop that can add up to systemic change.

Peer influence can be quite powerful. A recent article from the Yale School of Environment concluded that "the behavior of peers has a significant influence" on energy-related decisions such as installing rooftop solar or buying an EV. For example, residential rooftop solar tends to pop up in clusters: if one homeowner installs solar panels, neighbors who see the panels are more likely to do the same.[22]

Drawing on these ideas, I've applied three tests in deciding which personal actions to recommend in this chapter. I suggest you apply these in your mind as you read the six suggestions that follow.

First, feasibility. Am I able to do this? This test seems obvious, but even the most dedicated among us will not undertake actions that are too hard, too expensive, or make life significantly more difficult.

Second, impact. Will this matter if enough people do it? Using tote bags, recycling, composting, and planting native plants are all good for the environment. I like doing these things. But I know that even if everybody did them, it wouldn't slow climate change.

Third, spreadability. Can I help this action spread? Some things, like visible rooftop solar, spread naturally. Others, like divesting your savings from fossil fuels, can spread through conversations. It's not necessary to announce your move on Facebook, but it wouldn't hurt!

Six Personal Actions that Can Make a Difference

In the rest of this chapter, I describe six actions and show how they meet these tests. The six actions are:

- Stop Wasting Food and Eat Less Meat
- Drive Less: Walk, Bike and Take Public Transport More
- Move Your Money
- Install Rooftop Solar
- Upgrade Your Car to an EV
- Fly Less—or Not at All

I start with things you can do right away. If you haven't done so already, you can start today eliminating food waste and shifting to a plant-based diet. Same for driving less and walking and biking more. Moving your money—shifting your savings from banks and funds that invest in fossil fuels to those that do not—will take a little longer but can usually be done in a month or two. Installing rooftop solar and upgrading your car to an electric vehicle (EV) cost money but for many readers will still pass the feasibility test. My wife and I did the first five actions in the year it took me to write this book. I've left flying less or not at all for last, for reasons that you will see.

As you read each section, ask yourself 1) Am I willing and able to do this? 2) If enough people do this, will it make a difference? 3) Can I do this in a way that will cause the idea to spread?

Keep in mind, however, that even for actions that meet all three tests, it's crucial to not get stuck at personal actions. The fossil fuel industry and others who getting rich burning up the planet would like nothing better than for you to spend all your time and energy on personal actions. If you lack the time and energy to undertake personal action *and* become a climate activist, I suggest you skip this chapter and go to Chapter 5!

Want to do both? Let's get started!

Stop Wasting Food and Eat Less Meat

The World Resources Institute's massive, multi-year study, *Creating a Sustainable Food Future,* shows how to feed a global population on its way to 10 billion by 2050, while protecting ecosystems and reducing greenhouse gas emissions[23] The findings are distilled into a five-course menu. The goal of the first course is to slow the growth in demand for food, and the first two imperatives within it are to *stop wasting food* and to *eat less meat, especially beef*. These are simplest and most immediate things that you and your family can do to reduce your emissions.

Food production is responsible for about a quarter of global greenhouse gas emissions. Emissions occur at every stage: clearing forests and other carbon-absorbing ecosystems to make way for food production; plowing, which releases carbon from the soil; burning fossil fuels to plant, fertilize, tend, harvest, process and ship the food; and often using additional energy to keep food chilled or frozen. So it's a climate catastrophe—and a tragic waste in a world where one out of ten people are hungry—that one-quarter to one-third of the food produced each year never gets eaten—it is wasted between farm and fork.

While most food waste in developing countries happens before food reaches the market, due to pests or inadequate shipping or storage, in the United States and other high-income countries, most food waste occurs at the end of the trip. Food is produced, harvested, washed, processed and packaged, shipped to a store or delivered to a home or restaurant. It may then be thrown out without ever being prepared and served, or it may make it all the way to a plate and be stored as leftovers (using more energy to keep it cold), only to be tossed at the very last step of the journey. Result: the average American throws away about 400 pounds of food per year.

Reducing food waste is not only one of the simplest step you can take to cut your emissions, it also will save money on your grocery bill—or you might decide to spend the same amount but buy better quality ingredients. If enough people eliminate waste, it can slow emissions growth and reduce upward pressure on prices, helping to reduce global hunger. Were you told as a child to eat all the food on your plate because children elsewhere were hungry? Perhaps for you eliminating food waste will strike a chord, harkening back to one of the earliest ethical lessons you learned from your parents.

How to proceed? Stephanie Miller's *Zero Waste Living the 80/20 Way: The Busy Person's Guide to A Lighter Footprint* (2020) offers a handy guide to reducing your household food waste. If you are like me, you probably already do some of these things, while others will be new to you. All are simple and easy to incorporate into your household routine.

- **Plan Your Shopping:** If you don't buy it, you can't waste it. Miller cites a study by the World Wildlife Fund that one-out-of-three American families rarely or never take stock of the food they have on hand before buying more groceries. The solution: shop your cupboards and fridge as you make your grocery list—and always make a list!
- **Plan Your Meals:** Miller says she and her husband plan their meals weekly, meeting briefly on Saturday to jot down a menu that they post on the refrigerator. She suggests three plant-based meals, her husband suggests two meals, one night is set aside for leftovers and one night for eating out or ordering in.
- **Conduct a Daily Fridge Review:** Set a regular time that works for you—such as after breakfast or just before or after dinner. Identify anything that needs to be used right away. Move older foods towards the front and use see-through containers: what gets seen gets eaten.
- **Use It Up Cooking:** Have one or two go-to cooking techniques, such as a soup and a stir-fry, which allow you to "mix and match" ingredients. Use Google or a recipe app to find recipes that combine what you have on hand.
- **Use Your Freezer:** Can't use it now? Miller writes that she has been surprised at how many foods freeze well: soup stock, tomato paste, fresh berries. I freeze ripe bananas to toss in the blender with yogurt for a breakfast smoothie (slice before freezing). Don't forget to eat what you freeze. Following Miller's advice, I've begun checking my freezer when planning meals and conducting my daily fridge review.

Of all the foods we eat, meat and dairy emit the most greenhouse gasses. And the biggest emitters by far are the ruminants, mammals such as cattle and sheep, who have segmented stomachs and re-chew plant matter ("cud") to break it down for digestion. This involves a lot of burping which, contrary to what you might think, is a bigger source of cattle emissions than their farts.

The environmental and climate impact of meat production is astonishingly high. Meat production drives forest clearing in the Amazon, pollutes streams and rivers, consumes grain that could otherwise have nourished people, and emits both carbon dioxide and methane, a super pollutant with 20X the heating impact of carbon. If all the cattle in the world were a single country, their emissions would make them the third largest current greenhouse gas emitter in the world, after China and the United States.

From an emissions point of view, beef is by far the top offender. According to *Our World in Data,* producing a kilogram of beef emits 60 kilograms of greenhouse gasses (CO_2-equivalents), sixty times more than a kilogram of lentils, peas or beans. "Production of lamb and cheese both emit more than 20 kilograms CO_2-equivalents per kilogram. Poultry and pork have lower footprints but are still several times higher than most plant-based foods, at 6 and 7 kg CO_2-equivalents, respectively," writes Hannah Ritchie, head of research at *Our World in Data.*[24]

Shifting to a plant-based diet—one based around legumes, whole grains, fruits and vegetables—doesn't require becoming a vegan or vegetarian. Meatless Mondays, a national movement, is a great way to begin. You may find that your preference for plant-based foods grows over time, and that meat goes from being a routine part of your diet to an occasional treat. Many older people find that their desire for meat fades as they grow older. You may eventually find that you give up meat altogether.

Of course, eating less meat also offers important health benefits. According to the Mayo Clinic, people who eat red meat are at an increased risk of death from heart disease, stroke, and diabetes. Eating processed meats, such as bacon, cold cuts, and sausage further increases the risk of these diseases.

As with other personal actions, don't underestimate the social impact of your choices. Humans are social creatures, and we adjust our behavior based on what we see others do. Offering a plant-based meal next time you cook for family or friends, or selecting one next time you dine out, may have more influence on others than you realize. Conversations about food choices can provide an opening for discussing the climate emergency, and why it's important that all of us take steps to address it.

Drive Less: Walk, Bike & Take Public Transport More

You have heard the advice many times—drive less, walk, bike and take public transportation more. It's good for our health and good for the planet. It's true: reducing the miles traveled, and selecting low-carbon options whenever possible, is one of the simplest and most effective things you can do to reduce your emissions. And, as we have seen for cycling, transportation choices are spreadable.

America's appalling lack of public transport means that surface transportation is the largest source of carbon emissions in the United States, accounting for almost a third of emissions. Passenger cars and so-called light-duty trucks (SUVs, minivans, and pickup trucks) account for half of this pollution.

Burning a gallon of gasoline pumps about 20 pounds of CO_2 into the atmosphere. This may seem impossible since a gallon of gas weighs only about 6.3 pounds. But most of the weight of the CO_2 doesn't come from the gasoline itself but from oxygen in the air. Gasoline is a combination of hydrogen and carbon. When it burns, these elements separate; the hydrogen bonds with oxygen to form water (H_2O), and carbon bonds with oxygen to form carbon dioxide (CO_2).

The simplest way to emit less carbon pollution is to travel less, an easy option for boomers as we retire. No more commute! The COVID-catalyzed surge in teleworking tools for paid and volunteer work has greatly reduced the need for routine travel. For the trips that remain, walking, biking, using public transit, and—when you must drive—combining multiple errands into a single trip, can help reduce emissions, congestion and U.S. demand for oil. Other public benefits of reduced car travel include reducing noise, air pollution, and the destruction of wetlands and other open space through road widening projects.

As with eating less meat, walking more is good for you as well as for the planet. The Mayo Clinic says that walking can lead to higher level of fitness and health. Benefits include managing blood pressure, reducing the risk of diabetes, managing weight, reducing the risk of a heart attack, and managing stress and "boosting your spirits."[25]

Many people find it easier to incorporate walking into their daily routine if they combine it with errands. Are there nearby destinations you can walk to instead of driving, perhaps a library, coffee shop, or a grocery store? Would you be more likely to do it if you bring a backpack or a wheeled cart to carry home your purchases?

Bicycling is increasingly recognized as key to reducing transportation emissions. A recent European study collected travel data from about 4,000 people living in seven cities for two years, then calculated the carbon footprint for each trip. People who cycled daily had 84% lower carbon emissions from their daily travel than those who didn't. Here in the United States, the reduced emissions benefits might be smaller—after all, our public transportation and bike infrastructure are far behind Europe—but they would trend in the same direction.

Biking is getting safer and easier all the time. You needn't be a MAMIL (Middle-Aged Male in Lycra) to join in the fun. Many American communities are adding protected bike lanes, and many local bicycle associations offer coaching and group rides for new and inexperienced riders. If you need a little extra help with uphill climbs, consider an e-bike. Prices are falling and the range of e-bike models is growing.

Using public transportation is one of the most effective actions you can take to conserve energy and reduce emissions. One person who switches from commuting 20 miles alone by car to using existing public transportation would reduce their annual CO_2 emissions by 48,000 pounds a year, equal to a 10% cut in the emissions of a typical two-adult, two-car household.[26]

Moving into retirement or cutting back on work hours? Explore the public transportation options near you. While public transportation in the United States lags far behind other high-income countries, you may be surprised to discover options you weren't aware of, and that your new schedule makes it easier to use public transportation than when you were working full time.

Cutting back on car travel can make a significant difference. If all drivers in the United States drove 10% less, it would eliminate about 110 million metric tons of carbon dioxide emissions, the same as taking 28 coal-fired power plants offline for a year. Each driver would, on average, need to cut about 1,350 miles per year. For many of us that's an attainable target. Check how far you drove last year, set a 10% reduction target, and tell others, just as you might with some other personal goal. Sharing your intentions can serve as a commitment mechanism and can encourage others to do the same.

Of course, many of us will have no choice but to continue to rely on cars. Those of us who do will want to seriously consider making our next car an EV. I explain why a bit further below. But first I suggest you move your money. Why? We are starting with the easy, cost-free actions. Divesting from fossil fuels costs nothing and may even increase your wealth, while buying an EV is a significant expenditure. Read on to learn more about both.

Move Your Money

If your bank lends to fossil fuel companies, or your retirement savings are held in funds that include fossil fuel stocks, you own fossil fuels. You own the lies and the secret lobbying to block climate action. You own the pollution that causes cancer and asthma in the poor communities. You own the oil spills; the pipeline leaks; and the offshore blowouts that kill birds and fish and destroy livelihoods. You own these things because you are helping to provide these planet-killing companies with capital. By supporting financial institutions that bankroll them, you are implicitly endorsing their actions and giving them social license to operate.

As members of the generation that was mostly asleep at the switch when climate change went from being a distant future threat to an imminent planetary catastrophe, we boomers who understand the climate threat must put our money where our mouths are. We must move our money from the banks and funds that bankroll fossil fuel companies to those that are helping to hasten the shift to a low-carbon economy.

If enough of us move our money, we can make a difference. Th!rdAct, the movement of experienced Americans founded by Bill McKibben, cites research showing that 70% of America's financial assets—things like stocks, bonds, and bank accounts—are owned by boomers and our elders, the Silent Generation, compared to just 5% owned by millennials.

Th!rdAct and other groups campaigning to end fossil fuel finance cite the annual *Banking on Climate Chaos* report authored by a coalition of environmental NGOs including Rainforest Action Network, Indigenous Environmental Network, Oil Change International, and the Sierra Club. The 2022 report, endorsed by more than 500 organizations in more than 50 countries, ranks the fossil fuel finance (lending and underwriting of debt and equity) for the world's 60 biggest banks.[27]

It names a "Dirty Dozen" banks that together account for more than half of fossil fuel finance since the Paris Agreement was signed in 2016. The four worst—JPMorgan Chase, Citi, Wells Fargo and Bank of America—provided more than a trillion dollars in recent years, a quarter of fossil fuel financing.

Th!rdAct's national campaign to encourage older Americans to move our money from these four big climate-bad banks offers guidance on how to proceed, including a list of better bank and credit card options. Many credit unions and community banks, for example, provide similar services to the four big climate-bad banks but do not invest in fossil fuels.

THIS is What We Did, another organization that helps older Americans act on climate, supports the Th!rdAct campaign with additional guidance for selecting a climate-friendly financial institution, draft letters to bank CEOs telling them why you are moving your money, and step-by-step checklists. THIS also offers free weekly coaching sessions for groups of five or more. At the end of the four weeks, THIS promises that you will be ready to move your money.

Ensuring that your retirement savings and other investments are fossil free takes more work than switching banks. But it needn't cost you anything and will reduce your long-term risk, since the days of fossil fuel are numbered. Oil company stock prices are based on their proven reserves, on the assumption that these will eventually be burned—an unlikely prospect, since burning them would cook the planet. Divesting now will protect your savings when enough people realize this truth and the carbon bubble bursts.

Savvy investors have seen this already. "Over the last decade, fossil fuels have lagged the general economy in profitability," Tom Sanzillo, director of financial analysis at the Institute for Energy Economics and Financial Analysis, said in late 2021 at the launch of a report showing that some $39 trillion in assets had committed to some form of fossil fuel divestment. "For decades, oil and gas drove the economy and the financial markets. That is no longer true. There may be short-term stock price increases, but long-term value creation is decidedly negative... The oil and gas sector once commanded 28% of the Standard & Poor's 500 index. Today, it is less than 3%... The shift is titanic and it is poorly understood."

Although Russia's invasion of Ukraine rattled energy markets—and dramatically increased profits and share prices for oil companies—Sanzillo's prediction of short-term price fluctuations amid long-term declines has held up well. In late 2022, the industry leader, ExxonMobil, the biggest U.S. oil company, had a market capitalization of just over $400 billion, well short of the $575 billion it was worth in 2007.

There are two paths for ensuring that your retirement savings and other investments are fossil-fuel free: do it yourself or get an expert to help you. If you currently manage your money yourself, it makes sense to start with the DIY approach and see how far it takes you. Sometimes the first path can lead you to the second.

That's what happened with our family. Like a lot of busy, working boomers, my wife and I followed the standard advice to invest our retirement savings in low-cost index funds, in our case Vanguard's S&P 500. We worked hard, saved, and didn't pay a lot of attention to these accounts. When we decided to divest from fossil fuels, I contacted Vanguard. The company assigned us an "investment advisor" who recommended so-called ESG Funds, funds that purported to take into account a company's performance on three sets of indicators: environment (e.g., climate, air and water pollution), social (e.g., wage policies) and governance (e.g., board diversity) indicators.

We soon learned that ESG is the investment equivalent of the Wild West. There are no standard definitions, so while some ESG funds are an effective way to divest from fossil fuels, others are little more than standard mutual funds with an ESG label to justify higher fees. Sorting through the differences was going to take a lot of work. We also gradually realized we needed more help with our finances as we approached retirement, sorting through Medicare options, insurance, refinancing our mortgage, and simply consolidating retirement accounts. We had reached the point where we needed a Certified Financial Planner (CFP).

We asked around and discovered that some of our friends worked with financial planners who shared their commitment to social justice. We interviewed several and selected one, Jason Howell, whose values aligned with ours. Howell advocates a wholistic approach to managing money, which he describes in a book: *The Joy of Financial Planning: 7 Strategies for Transforming Your Finances and Reclaiming Your American Dream.* Long story short, he and his business partner helped us to put our affairs in order and in the process to divest from fossil fuels.

Perhaps you already have a financial planner? Tell them you want to bring your holdings in line with your values and divest from fossil fuels. They may already have the skills and knowledge to help you, or they may be hearing similar requests from other clients and be open to developing such expertise. If your financial advisor is willing to travel this path with you, these conversations can help the idea to spread. If not, it may be time to find somebody else.

Whether you are thinking about changing advisors or looking for a financial advisor for the first time, it's important to pick one who can help you eliminate fossil fuels from your portfolio while protecting your wealth. Green America, a nonprofit organization that advocates for environmental and social economic justice, offers a list of financial advisors who help people align their investments with their values. Look for an advisor who is a Chartered SRI Counselor (CSRIC) in addition to being a Certified Financial Planner (CFP).

Perhaps your holdings are small enough or simple enough that it doesn't make sense to work with an advisor. Or maybe you have a lot of experience managing your money and prefer to handle it yourself. Several organizations offer step-by-step guides to Socially Responsible Investing (SRI). In November 2021, Kiplinger, the Washington, DC-based publisher of business forecasts and personal finance advice, ranked As You Sow as the best of them.[28] According to Kiplinger:

> As a national non-profit for shareholder advocacy, As You Sow offers one of the most comprehensive SRI/ESG tools available to investors. On its site InvestYourValues.org, investors can evaluate the social responsibility of 3,000 mutual funds and ETFs, from their fossil fuel impact to gender equality and racial justice through the companies' relationships with private prisons.

Andrew Behar, the CEO of As You Sow, says one way to use their tools is to start with a list of the stocks or funds you currently hold. Go to *InvestYourValues.org* and select one of the financial screens (filters). This will take you to a new page where you can type the name, ticker or manager of a fund into the search box. The result will show how that fund scores on an A-to-F scale for the financial screen you chose. Scrolling down will reveal the fund's score on all the other metrics As You Sow provides.

If it is not aligned with your values, look for something better by clicking "Search Funds" in the navigation bar. This lets you view all funds rated by As You Sow. Then use the "Search Settings" option to refine your results by fund family, category, performance or minimum grade on any given screen.

As You Sow's newest campaign highlights the responsible investment options—or lack thereof—in many 401(k) plans. If you hold money in an employer-sponsored 401(k) that lacks fossil-free options, the site offers tools to help you engage with your employer and plan sponsor, including tips for organizing your colleagues and a draft letter you can send to your 401(k) provider.

As with other personal actions, you can magnify your impact by helping the idea to spread. This is a little harder with financial planning than with some other actions. For most of us, money is a private matter. We aren't accustomed to discussing our financial decisions with friends and neighbors. Start by sharing your journey with those you might normally discuss finances with, such as your family members. We told our adult children and were surprised to discover that their early, modest retirement accounts are already invested fossil-free. And, of course, you can share the broad outlines without getting into particulars. Here's a simple conversation starter: "We are more and more concerned about climate change and recently decided to divest our retirement savings from fossil fuels. I'm new to this. Have you done anything like this?"

Upgrade Your Car to an EV

Let's say you have already taken the steps described in "Drive Less: Walk, Bike & Take Transport More." Many of us will still need a car, especially those of us who live in suburbs and towns where distances are long and public transport is in short supply. If you are in such a situation, should your next car be an EV? Almost certainly, the answer is yes.

Meeting our country's Paris Agreement commitment of cutting emissions in half by 2030 and being carbon neutral by 2050 will require rapid transformation of the U.S. auto fleet from gasoline-burning internal combustion engine vehicles (aka ICEs) to EVs. To help meet these goals, President Biden issued an executive order in late 2021 targeting sales of new EVs to hit 50% by 2030.[29] Automakers and labor unions are increasingly on board. Biden was joined by representatives of Ford, GM, Stellantis (a European manufacturer created from the merger of Fiat Chrysler and the French PSA Group), and the United Auto Workers (UAW).

Environmental leaders applauded Biden's goal but warned it wouldn't stand a chance without supportive policies and regulations. Swapping your ICE for an EV can help speed the transition, making the policy and regulatory lift just a little less demanding—especially if you become an EV convert and talk up your smart choice with friends and relatives!

Norway, the world's leader in battery-powered EV adoption, shows how quickly the transition can happen. In 2021, more than 65% of new car purchases in Norway were battery powered. Rather than offering subsidies, Norway revamped its tax structure, exempting EVs from a 25% tax on gasoline vehicles. The country also invested in a first-class charging structure, including incentives for housing cooperatives to install public stations. For taxis, Norway rolled out wireless induction charging, installing plates in the roadways at taxi stands.

The United States has a long way to go to catch up with Norway but is headed in the right direction. In 2022, U.S. EV sales grew 62% from the year before, to 6% of all new passenger vehicles. Tax incentives for new and used EVs in the 2022 Inflation Reduction Act (IRA) will bring EV prices close to those of gasoline-powered cars, further boosting sales.

Charging stations are multiplying, and the distance you can travel on a single charge is increasing. Further range improvements are expected: in June 2022 BMW announced it is testing a new battery-powered SUV that can go 600 miles on a single charge. The new battery uses two kinds of chemistry, one for power delivery and one for energy storage, and requires significantly less minerals, such as lithium, nickel and cobalt, reducing environmental impact.[30]

Does an EV really reduce emissions? Absolutely. Studies show that an EV is already the lowest emission choice, even in areas where the grid includes a lot of coal, since EVs are highly efficient and at least some of the electricity is generated by other means. Where I live in Virginia, for example, renewables were just 6% of utility scale power in 2020, but nuclear power, another zero-emissions source, accounted for nearly a third of the power supply, so my recently purchased EV is much cleaner than the car it replaced.

Moreover, an EV you buy this year will become less polluting each year, as the electric grid gets cleaner. David Roberts, who writes on energy and climate at Vox, explains that an EV—or any other electrical device, such as a heat pump that replaces a methane (aka "natural gas") furnace—will get cleaner throughout its life, as the grid gets cleaner with the closure of coal plants and the addition of carbon-free renewables.

Saul Griffith, author of *Electrify: An Optimist's Playbook for Our Clean Energy Future*, urges readers to help speed the transition by replacing machines that run on fossil fuels with those that run on electricity. When your methane-powered furnace, water heater, clothes drier, or cooktop needs replacing, he writes, replace these with the high-efficiency electrical alternatives, such as heat pumps (for furnaces) and induction cooktops (for methane-fueled ranges). It makes sense for most of us to start with the biggest machine we own, our car.

Plus, EVs are fun to drive. My wife and I discovered this in 2021 when we gave our 1997 Toyota Echo to our son, who lives in New Orleans, to use as a hurricane escape vehicle, and bought a two-year old Hyundai Kona Electric, which we selected for its long range—380 miles when fully charged.

Soon after, we drove from our home in Arlington, Virginia, to Lake Ontario, where we had rented a lakeside cottage for a summer get away. Finding chargers wasn't always easy—we had a couple of anxiety-inducing close calls. We quickly learned that on long trips it makes sense to charge fully when passing through a big town or city, whether you are low on power or not, since, for now at least, it might be a long way to the next charging station. Still, we find that benefits are well worth the extra bit of planning: the car accelerates quickly, smoothly and quietly; and there is zero tailpipe exhaust. Plus, new EV mapping apps are making it much easier to trace a charging route to your destination. Did I mention that it's fun to drive?

For now, EVs cost more to buy than gasoline-powered cars, but the gap is narrowing. In 2021 the most popular non-truck vehicle was the Toyota RAV4, with a starting price of $26,350, compared to the Hyundai Kona Electric with a starting price of $34,000. Depending on your income, federal subsidies may cover much of the $7,650 difference. The 2022 Inflation Reduction Act includes a $7,500 tax credit for new EVs, and $4,000 for used EVs. The law imposed new income caps but also made the subsidies easier to use, by enabling buyers to claim them at the time of purchase, rather than waiting until they file their taxes.

When it comes to the full cost of ownership over time, EVs are much cheaper due to much lower fuel and maintenance costs. If a Toyota RAV4 and a Hyundai Kona EV are each driven 14,000 miles per year, fuel costs for the EV are about half those for the ICE, for an annual savings of about $700. And since EVs have no engine—only an electric motor with few moving parts—there's much less wear and tear. EV owners can look forward to saving an estimated $4,600 in maintenance costs over the life of the vehicle.

The biggest potential maintenance cost for EVs is to replace a battery, a major cost roughly comparable to replacing the engine in an ICE. But batteries are getting better all the time, and some EV's come with an eight-year or 100,000 mile warranty. Moreover, new evidence suggests that EV batteries are lasting much longer than predicted—or warrantied. Some experts now say that most EV batteries will still have about 80% of their capacity when the car they are in is finished, making them useful as back-up storage for solar arrays.[31]

What about charging? Every EV comes with a cord that plugs directly into any standard wall outlet for so-called Level 1 charging using ordinary 120 volt household current. For most people, charging with a Level 1 charger overnight will give them enough power for the day ahead. Level 2 chargers require 240 volts, the level of power that runs a typical clothes dryer. Newer homes with garages often come equipped with such outlets. An older house like mine will need an electrician to install one, an additional cost that can sometimes be offset by tax rebates.

Because Level 2 chargers have more juice, they charge faster. A Level 1 charger provides about 4 miles of range per hour of charge, compared to 32 miles per hour of charge for a Level 2 charger. A Level 3 charger, available in places like Walmart parking lots and public parking garages, will give you 3–20 miles of additional range for every minute of charging. And charging stations are set to proliferate: the Bipartisan Infrastructure Bill passed in November 2021 includes $7.5 billion to build a nationwide network of 500,000 EV chargers.

EVs, like bicycling and installing rooftop solar, are highly visible and therefore spread easily. The more people who buy EVs, the more others will come to see this as an appealing alternative. You may not have to do much to spread the idea. Friends and neighbors often ask us how we like our EV. We are happy to tell them how delighted we are with our purchase.

Install Rooftop Solar

If you live in a single-family home, installing rooftop solar panels can greatly reduce—and in some cases, eliminate—

emissions associated with your home power consumption. Solar energy is reliable, costs are falling, and tax incentives and a variety of financing plans make it easier than ever to go solar. By installing solar, you can help to accelerate this transition to clean, renewable energy.

If you live in an apartment or condominium, you may still be able to benefit from—and help to accelerate the growth of—renewable energy through a community solar program, an option I discuss briefly below.

Electric power accounts for about a quarter of the country's emissions of heat-trapping gasses. Fossil fuels—mostly coal and methane gas—provide about 60% of U.S. electricity, but account for nearly all the emissions. Each rooftop solar installation makes the country's power supply just a little bit cleaner and hastens the day when the last coal-fired power plant will finally close.

We have seen that rooftop solar spreads by example—people are much more likely to go solar when they see that their neighbors do it. Solar also offers the same virtuous circle as with bicycles: the more people who install it, the greater the pressure on regulators and public utilities to streamline the permitting process and make other changes to make solar easier and more worthwhile, which in turn leads to more solar. By installing solar, you strengthen this virtuous circle.

With rooftop solar, the costs are upfront and the benefits—free electricity for years to come—accrue slowly. A typical household installation costs $15,000 to $40,000, depending on the size of your house, a cost that can be partly offset by a 30% federal tax deduction, plus incentives from some state and local governments. Payback periods—the time it takes to break even and start saving money—typically range from seven to ten years, depending on the size of the installation and the cost of the electricity from the grid.

For some of us this may be longer than we expect to remain in our homes—and might even begin to bump up against our life expectancy. Even though it will take several years to realize the financial benefits, my wife and I decided to go solar anyway, for two reasons. First, we believe it's the right thing to do to address the climate emergency. Second, while we expect to remain in our home for the foreseeable future, if we sell before we break even, homes with rooftop solar tend to sell for more than comparable homes without, so we may recoup much our investment anyway.

So, how to get started? Several online tools offer quick estimates based on your address and your monthly electricity bill; some share that information with installation companies who will contact you to pitch their services. Of the more than 10,000 solar installation companies in the United States, a handful have national reach, most are small and medium-sized firms. Because regulatory regimes and subsidies differ by locality, you want a company with plenty of experience in your area.

Ask around your neighborhood. If you see a house with solar, don't hesitate to knock and ask for advice. Nearly everybody who has installed solar has done so to help hasten the energy transition and will be happy to share their experience. Ask your social media networks—Facebook or the hyper-local NextDoor—for recommendations. Sharing your interest in going solar can spark conversations that spread the idea.

As you identify firms, check their websites and references. Most offer a free consultation that leads to a project proposal showing cost comparisons of solar versus no solar. Many will meet with you, at your home or online, to show a Google Earth image of your rooftop and explain where they propose to place the panels based on shade coverage and the angles of your roof exposures. Watching a solar expert "drag and drop" solar panels on your roof is a fun way to imagine what it would be like to go solar. There is no obligation to proceed.

As with any major home expenditure, get three bids, then narrow the field based on cost, quality, and reputation. Look closely at the guarantees. While solar systems are highly reliable and usually require little maintenance, I liked the company I hired—Ipsun Solar—because they hold the warranties on the system components, providing buyers a single overarching guarantee, backed by insurance from Swiss RE, which would pick up the tab if Ipsun went out of business. They monitor the system remotely and fix anything that goes wrong.

When selecting an installer, look for a "Certified B Corporation." Certified B Corps are value-based businesses that are legally required to consider the impact of their decisions on their workers, customers, suppliers, community, and the environment—part of the growing movement to harness the power of business as a force for good. BLab, a not-for-profit that provides certification and advances the B Corp movement, lists 20 solar installers; you can filter these by size, location, years of certification, and other qualities, such as minority or woman-owned businesses. Ipsun, the company I hired, is a B Corp that has an active presence in Virginia's capital, working to remove regulatory obstacles inhibiting the growth of solar.

If you live in a multifamily building, community solar may be an option. These programs also make solar accessible to people with low-to-moderate incomes, renters, and households without adequate roof space or solar exposure. Rather than installing panels on their home, community solar users subscribe to a shared system, often located within their community. The Department of Energy's National Community Solar Partnership offers information to get you started.[32]

Fly Less—Or Not at All

I have left "Fly Less—Or Not at All" for last because, for me, it is the most difficult.

I confess, I love to travel. When my wife and I we were in our 20s, we lead some of the first U.S. commercial tours to China, which involved flying trans-Pacific twice a month or more, plus flights between China and Japan and countless flights within China. Later, living and working in Asia, and then for the World Bank and think tanks in Washington, DC, I flew for business and pleasure, seeing things that I never could have otherwise, from Petra, in Jordan; to site of the Treblinka Concentration Camp outside of Warsaw; to a school in the foothills of the Himalayas where my mother and aunt boarded for a year during World War II.

If there is climate karma—payback in a future life for bad deeds in this life—I'm in serious trouble. Even as I've become aware of the climate emergency, I've continued to fly—although less than before. Pre-COVID, my wife and I flew to London for my first visit to Scotland, my ancestral homeland. I was proud that we did the land-based part of our trip via trains, buses, ferries and taxis. Of course, whatever emissions we avoided by not renting a car were tiny compared to those of our trans-Atlantic roundtrip flights. About once a year, we fly from our home in Northern Virginia to visit our grown children in San Francisco and New Orleans.

Whether this poses an ethical problem depends on how we think about aviation emissions. Civilian aircraft account for only about 2.5% of total global CO_2 emissions, so they might not seem to be a big deal. But that's only because so few people fly. A 2020 study estimated that eight out of ten people in the world have never been on an airplane and that just 1% of the world's population is responsible for more than half of aviation emissions.[33]

"The air industry and its lobby are keen to portray air travel as a normality, when really one ought to view this as a very unevenly distributed privilege to which few have access," said the study author Stefan Gössling, a professor at Western Norway Research Institute.

For those of us who fly—a group that likely includes nearly everybody reading this book—the atmosphere-harming emissions from our air travel are likely bigger than from anything else we do. This is morally problematic to say the least: air travel benefits those of us who fly, while imposing significant costs on people around the world who will never set foot on a plane.

Of course, the most effective way to reduce your aviation emissions is to fly less—or not at all. Teenage climate activist Greta Thunberg famously crossed the Atlantic in 2019 in a zero-carbon yacht that utilized wind, sun and hydro power. Thunberg, who introduced the world to the Swedish word *flygskam*—"flight shame"—is not alone in wanting to avoid air travel for ethical reasons. A study commissioned by the World Economic Forum (organizer of the jetsetter corporate gabfest known as Davos) found that one-in-seven "global consumers" said they would avoid flying to cut emissions, even if the alternative were less convenient or more expensive.[34]

Dan Castrigano, a former teacher who runs a climate organization from his home in Burlington, Vermont, is among the small but growing number of Americans who have sworn off flying. "There was this cognitive dissonance when I would fly," he told *The Guardian*. "I was teaching about climate to seventh and eighth graders, and I just kind of became embarrassed that I was flying to Europe for vacation." Eventually, he decided to give up flying altogether. Now he helps run Flight Free USA, which connects people who have pledged to avoid planes for a month, a year, or indefinitely. "It's extremely joyful not to fly," he says. "It's liberating."

For land-based travel, trains are the lowest-emissions alternative to flying. Although the U.S. rail network is embarrassingly inadequate—China has 22,000 miles of high-speed rail, the United States has none—serviceable passenger trains do ply some of the busiest U.S. travel corridors, like Washington, DC, to Boston. And the more people who ride them, the greater the demand for improved routes and service. So, take a train if you can and post highlights to your social media account, to inspire others to opt for the train.

If rail isn't an option, driving an EV or even a high-mileage gasoline-burning car produces fewer emissions than flying—and the more people in your car, the lower the per capita emissions. Driving long distances to minimize emissions is not as far-fetched as it may sound, especially for boomers who are retired or can control their work hours. My friends Elaine and Terry recently drove coast-to-coast in a newly purchased Volkswagen EV. Elaine is retired; Terry works as a Japanese-to-English translator and takes his work on the road.

"We wanted to visit friends and family in California but were uncomfortable with the emissions associated with flying," Elaine said. "It was also an experiment to see how hard it would be to keep the battery charged. With a little planning it turned out to be remarkably easy."

If you decide to fly, you can minimize your emissions by selecting non-stop flights, because take-off and landing are the most carbon-intensive part of a flight. Opting for economy class (so you take up less space) and packing light can also reduce your emissions, because the heavier an aircraft, the more fuel it consumes.

What about carbon offsets? The concept is simple: for each kilogram of CO_2 you emit, you pay somebody to not emit or to absorb from the atmosphere an equal amount of CO_2, for example, by protecting forests or planting trees. One group, Sustainable Travel International, provides a calculator that shows one person flying economy class from Washington, DC, to San Francisco emits about 1 ton of CO_2, which they offer to offset for just $17.28. Their website describes projects they support in the United States and abroad promising: "we carefully vet all of our carbon offset projects to ensure that we only support high quality projects that truly reduce CO_2 emissions and deliver meaningful impacts."[35]

Yet many people are uncomfortable with offsets. British author George Monbiot famously compared them to the sale of indulgences by the Catholic Church in the Middle Ages. "Just as in the 15th and 16th centuries you could sleep with your sister and kill and lie without fear of eternal damnation, today you can live exactly as you please as long as you give your ducats to one of the companies selling indulgences" he wrote in 2006. "By selling us a clean conscience, the offset companies are undermining the necessary political battle to tackle climate change at home. They are telling us that we don't need to be citizens; we need only be better consumers."[36]

Hostility to offsets isn't limited to climate activists. In 2021 United Airlines CEO Scott Kirby said that offsets were "a fig leaf" that enabled airline executives "to pretend that they have done the right thing for sustainability when they haven't made one wit of difference in the world." United, he said, planned to eliminate its emissions through new low- and zero-emissions jet fuel and direct air capture of CO_2. He pointed out that there isn't enough space in all the world to plant enough trees to absorb industrial society's emissions.[37]

What's a conscientious person to do? If you feel you must fly, for business, to visit loved ones, or perhaps even to fulfill a lifelong dream of travel to a distant land, I suggest you consider an alternative approach: use a reasonable estimate of the cost of the damage caused by your trip and donate an equivalent amount of money to an organization that is actively and effectively working for a rapid end to the fossil fuel era.

Economists have devised a measure called the *social cost of carbon* that estimates the damage caused around the world and into the future by one ton of carbon pollution.

Governments and companies use the social cost of carbon as a shadow price when conducting cost-benefit analysis. For example, in calculating the costs and benefits of a road widening project versus a new bus lane, the climate damage caused by the higher emissions from the additional cars traveling on the wider road can be calculated using the social cost of carbon and included when deciding between the two options.

The U.S. government first estimated the social cost of carbon during the Obama administration, setting it at $51 per ton, which many experts thought was too low. In early 2017, the National Academy of Sciences recommended a major update to make the calculation more transparent and scientifically sound. When Trump became president, he disbanded an interagency group working on an answer. His new team excluded damages abroad, slashing the social cost of carbon to between $1 and $7 per ton. Biden reinstated an interagency working group and in September 2022 the Environmental Protection Agency announced a new estimate of $190 per ton.[38]

I believe that calculating the damage caused by flying using a social cost of carbon of $190 per ton, then making a donation equal to that amount to an organization that is effectively fighting to end the fossil fuel era, is a better approach than buying voluntary carbon offsets at $17 per ton. That the social cost of carbon is a non-trivial amount is part of the reason: current offsets are too cheap to discourage flying. At $190 per ton—the amount of CO_2 emitted by one person on a U.S. coast-to-coast flight—the social cost of carbon will make potential travelers think twice: do I really need to take this trip?

While this doesn't entirely address Monbiot's allegation that paying to emit carbon is the equivalent of buying indulgences, it at least acknowledges the damage caused by flying and puts an equivalent amount of money towards a rapid transition to a low-carbon economy.

For a variety of reasons, I'm not yet ready to give up flying. For one thing, as my travel-loving wife approaches retirement, an announcement that I am no longer willing to fly would strain our marriage. So, I'm flying less and, when I do fly, I'm following the plan above.

For example, for a recent trip to California to visit family and participate in my 50th high school reunion, I used the Sustainable Travel International calculator to learn that my round-trip flights would result in the emissions of two tons of CO_2. At $190 per ton, the social cost carbon for my trip worked out to $380. That's a significant additional cost for me but still within my ability to pay. I donated that amount to Oil Change International, a small research and advocacy organization that punches above its weight in pushing for an end to the fossil fuel era. Alternatively, I could have given it to a bail fund supporting people blocking new fossil fuel infrastructure.

Not everybody will be willing or able to do this. A recent, rigorous study found that ordinary air travelers' willingness to pay for carbon offsets is just one Euro, a bit more than a dollar, per ton.[39] Still, I've been surprised to find how many of my friends say that they are open to the idea. If you have read this far, I hope you will consider it as a means for weighing your decisions about whether and when to fly in a new light.

Taking Action on the Six Actions

We have seen that the concept of "carbon footprint" was invented by British Petroleum to shift responsibility for the climate crisis from the fossil fuel companies to ordinary people, families like yours and mine. I suggest we reject this notion, while nonetheless taking steps to reduce our emissions in when we are willing and able to do so, when it would make a difference if everybody did it, and when we can do so in a way that causes the idea to spread.

In this chapter, we explored six actions that can potentially meet these tests: stop wasting food and eat less meat; drive less and walk, bike and take public transport more; move your money; swap your gas-burning car for an EV; install rooftop solar; and fly less or not at all. Perhaps you have done some of these things already and have been thinking about others. Take a moment now, before you turn the page, to decide which of these you will tackle next. Maybe you want to jot numbers in the margins for the order in which you will tackle them, or place stars next to one or two that you want to make a priority.

Remember to do them in a way that spreads—tell your friends what you are doing and why and ask them if they have ever considered doing something similar. But keep in mind that your role in averting climate catastrophe is more than the sum of your emissions and the choices you make as a consumer. Avoiding planetary catastrophe will require rapid systemic change, change that can only happen if enough of us become climate activists. We turn to this in the chapters that follow.

Chapter 4: Every Little Bit Helps, Right?

Action Checklist

- Stop wasting food and eat less meat
- Drive less: walk, bike and take public transport more
- Move your money
- Upgrade your car to an EV
- Install rooftop solar
- Fly less—or not at all
- Don't get stuck at personal action!

How can I work with others?

Working with others is crucial. Assess your strengths, activate your networks and seek out national organizations to find local connections. Bonus: have fun, make friends and bring new meaning into your life.

Chapter 5

How can I work with others?

As older Americans, we bring to the fight to avert climate catastrophe our knowledge and connections. We know ourselves, we know a lot about the fields we have worked in, and we know a lot of people. Your bundle of knowledge and connections is unique to you, a power that you alone can bring to the climate movement.

Tapping your knowledge and connections to work with others helps to build the power needed to achieve rapid, systemic change. It can also be deeply rewarding. I know from my own experience that facing the reality of a looming global catastrophe can cause anxiety, grief, even depression. Working with others helps to overcome these feelings, bringing renewed hope, courage, and joy. Some of the most fun I have had in recent years is planning and carrying out activities with new climate activist friends.

In stepping beyond the personal actions described in the previous chapter to work with others in one or more of the ways described in this and the chapters that follow, you will become a part of a growing community of passionate, creative, and determined climate boomers. Welcome to the club!

I begin this chapter with suggestions for taking stock of your strengths and networks. I then show how to use national organizations to find local connections, highlighting several that welcome boomer volunteers. The chapter ends with a look at three arenas where we can be especially effective: universities, pension funds, and professional networks.

Assess Your Strengths

You have a lot to offer! After all, you have spent the better part of a lifetime learning and acquiring skills. Some of these traits are so much a part of you that you may take them for granted. Because you are starting a new endeavor, it's a good time to take stock of what you can bring to the movement.

"Know thyself" is said to have been the first of three maxims inscribed in the forecourt of the Temple of Apollo at Delphi in ancient Greece. Three millennia later, it's an excellent starting point for charting your path to climate activism.

You have reached a point in life where you know yourself well. You know what you like and what you don't like, which tasks you excel at and which you merely muddle through. As a climate activist, you may find that you use familiar strengths in a new way. Or you may discover skills you didn't use in your day job that are very much in demand as you explore a new role as a climate activist.

Think back on your own life: your experience with family, friends, and in the workplace. What tasks give you joy? When have you felt effective and rewarded in working with other people? What ways of thinking and being could you draw upon in pushing for climate action?

During your working life, you may have participated in a self-assessment to inform your career choices and help you work better with others. These tools assess things like values, interests, personality, and skills and aptitude. They include Myers-Briggs, Keirsey Temperament Sorter, Big Five Personality Assessment, and Holland Code. If you used one of these and found it useful, you might want to do it again. Many are available online, some for free and others for a modest fee. All yield a report you can use to reflect on what you can bring to this new stage in your life.

My favorite is the StandOut Strengths Assessment created in 2011 by human resources guru Marcus Buckingham, who now leads the ADP Research Institute's People and Performance Studies. You might remember ADP as a timesheet and payroll processing platform. Starting from these back-office roots, it has grown to offer a full suite of tools and a theoretical framework to help organizations become more effective by enabling employees to identify and use their strengths.

To help people cope with the disruptions of the COVID pandemic, the ADP Research Institute made the Strengths Assessment available for free online. It takes just 15 minutes to complete: it relies on top-of-mind reactions, so it's timed. At its core are nine strengths—Advisor, Connector, Creator, Equalizer, Influencer, Pioneer, Provider, Stimulator, Teacher. After you complete the assessment, you'll immediately receive a 14-page results report that identifies your two biggest strengths and explains how you can apply these to achieve success.[40]

When I took StandOut while working as vice president for communications at the World Resources Institute, I learned that my top two strengths are Connector and Creator. I wasn't surprised by the Creator moniker. I've always considered myself a creative person—I love to garden, cook, write. Heck, as a teenager I made pottery and copper jewelry.

The StandOut description of creators also resonated with me: "You like to find the patterns in life... For you there's nothing quite as thrilling as finding a pattern that can explain why things play out the way they do." Yup, nailed it. For me, part of the fun of writing this book is seeking and naming patterns.

What surprised me, although in hindsight it shouldn't have, is that StandOut identified my biggest strength as Connector. "You love to introduce people and are excited about the possibilities of these connections," I read. "You are a naturally inquisitive person, always asking questions to learn about others. The world for you is a network of people to be linked to create and accomplish better things."

Wow. To paraphrase the Scarecrow in *The Wizard of Oz*, that was me all over the place! Yet I never knew this about myself. I assumed everybody liked making connections as much as I do. Now I saw my desire and ability to connect people as a strength. Knowing this, I began to seek opportunities to connect people to accomplish better things.

For me, writing this book combines creativity—identifying and naming patterns—with making connections. Nothing would make me happier than if you use this book to discover connections to the individuals, groups and networks that will make you a significant actor in the most important movement in the history of humankind.

You could conduct your self-assessment without a tool. After all, you know yourself well by now. But why not give it a try? You might be surprised, as I was, to identify a strength you have been exercising without knowing it. Or you may gain new vocabulary to help you think through and articulate your new role as a climate boomer.

Ready to try? Can you spare an hour right now to take the 15-minute strength assessment and read and reflect on the report you will receive? If so, do it now! Here's the link to StandOut (www.MarcusBuckingham.com/test/) or use any other self-assessment tool you like.

If you can't do it now, I suggest you commit to setting aside an hour this week. Go ahead, say it outloud: "I commit to taking the StandOut Strength Assessment this week." Decide when, put it on your calendar, send yourself an email, or just note your promise in the margins of this book.

If you use StandOut, you will be in good company. Buckingham says that 750,000 people have taken the assessment and that StandOut is used by thousands of organizations around the world. Perhaps, like me, you will discover something new about yourself!

Take Stock of Your Networks

Next, map your networks. Since climate change affects everything and everybody, there's likely a climate angle to almost any organization that you already belong to or support. Examples: alumni, homeowner, neighborhood, and professional associations; school boards, local and state governments, and religious institutions, to name just some of them.

There are plenty of sophisticated tools for mapping social networks. For our purposes, simple is best. Take a piece of paper and a handful of markers or color pencils and draw a simple diagram of your networks. Put yourself in the center and draw circles, clouds, or other shapes clustered around you.

You will probably draw bigger circles for larger networks or for networks that are more important to you—that's good. Now connect yourself to the clusters with lines of various weights, using thick lines for stronger relationships and lighter or dotted lines for less strong ties. Are some clusters linked to others, for example, does your bike riding club include a lot of folks from your neighborhood (that's the case for mine). Consider adding lines to show these connections.

As you think of more clusters and connections, you may want to start over, perhaps drawing the map several times. You may want to group some of the clusters—schools you attended, or places you worked, for example. It's OK that it's messy! What matters is surfacing what's already in your mind.

Because climate change affects everything and everybody, each of the clusters will include some people who share your concern about climate change. In many cases—especially with the larger groups—you may find that there are already sub-groups working on climate, perhaps promoting one of the personal actions in the last chapter or advocating for various local, state, or national policy initiatives.

Ask around within your existing networks, sharing what you have been learning from this book and other sources. Be on the lookout for people who share your concerns. You might want to ask a life partner or a close friend to do their own self-assessment and map their networks, so that the two of you can compare notes. Saying it out loud will help make it real and each of you will likely have insights about the other.

Take notes on what you find, whether in digital form or perhaps in a new notebook you buy for this purpose. Regardless of how you keep these new records, hold them in your mind as you read the examples below of ways to act in community. Watch for overlaps between your world and the growing world of climate activism, searching for the niche where you will feel inspired to make your unique contribution.

Use National Orgs to Find Local Contacts

One of the easiest ways to find local connections to add to your network is to start with a national organization that offers education, training and networking opportunities. Many have local affiliates that you can find on their websites; others will connect you through an online sign-up form. Many also offer monthly video calls to welcome new members. Where local affiliates are lacking, the larger organizations offer toolkits and advice to start your own.

The sudden proliferation of Zoom and other virtual meeting tools since the start of the COVID-19 pandemic makes it easier than ever to attend trainings and meet like-minded folks. Online meetings are a convenient entry point for discovering people who share your concern and want to do something. Whenever possible, add an in-person dimension to these new friendships, meeting over lunch or coffee, or at an in-person training or action.

Below are seven organizations that can help you along the path to becoming a climate boomer. Shop around! Visit their websites and see which resonate for you. Sign up for their free email newsletters. Attend the monthly welcome meetings. You will soon get a sense of which of these organizations—and the two that I describe at greater length below—might be a good fit for you.

Chesapeake Climate Action Network (CCAN). Founded in 2002 by Mike Tidwell, a veteran journalist and author of books about the impact of climate change on coastal communities, CCAN organizes volunteers in Maryland, Washington DC, Virginia, and beyond to fight for bold, just climate solutions in the Chesapeake region, across the country, and around the world.

Climate Reality Project. Climate Reality Project carries forward the work and messages of Al Gore's pathbreaking documentary, *An Inconvenient Truth.* It offers in-person leadership corps trainings and encourages supporters to join one of more than local 140 chapters. According to their website: "whether you're a lifelong environmentalist or a new activist just starting out, there's a place for you in your local Climate Reality chapter."

Elders Climate Action. Founded in 2014, Elders Climate Action is the first national organization focused on mobilizing older Americans for climate justice. It organizes letter writing campaigns, rallies, non-partisan get-out-the vote efforts, and other activities. The group's "Letters to Loved Ones in 2050" project posts letters in which elders promise to do all they can to preserve a livable planet. An online directory lists state and local chapters.

Moms Clean Air Force. With chapters in 26 states, Moms Clean Airforce advocates for clean air for kids and against climate change. Recent campaigns include promoting electric school buses and supporting national climate legislation. Designed for busy moms (and dads), it focuses on letter writing campaigns. A map on the website makes it easy to find state-level chapters.

Sierra Club. One of the nation's oldest and largest environmental organizations, the Sierra Club has more than

60 chapters nationwide. While it's best known for outdoor recreation, it's Beyond Coal Campaign aims to close all U.S. coal-fired power plants and to stop new methane gas connections. Their website lists nearby chapters, training opportunities, and workshops.

THIS Is What We Did. A scrappy, San Francisco Bay Area organization focused on boomers, THIS answers the question that younger Americans will surely ask us: "What did you do to fight climate change?" It offers a climate literacy quiz, free classes on effective climate conversations, and coaching support for moving your money out of fossil fuels.

350.org. Founded by Bill McKibben and students at Middlebury College in 2007, 350.org pioneered digitally enabled climate organizing. As I noted in Chapter 1, it takes its name from the 350 parts per million (ppm) of CO_2 in the atmosphere that is the upper limit for avoiding the worst effects of climate change. The site offers resources for organizers and a map for finding a local chapter.

Unleash Boomer Power with Th!rdAct

McKibben, Th!rdAct's founder, has written more than a dozen books about climate change and is arguably America's best-known climate activist. His book *The End of Nature* (1989) was the first book on climate change to reach a large, general audience. After stepping down from leadership of 350.org in 2014, he published widely on climate, including regular columns for *The New Yorker*. In 2021 he announced that Th!rdAct would organize "experienced Americans" to end reliance on fossil fuels and protect democracy. Within 24 hours, thousands of people had signed up.

From the start, Th!rdAct has displayed many of the hallmarks of the early years of 350.org, strengths derived from McKibben's deft touch as a leader and communicator, including a compelling articulation of the organization's goals and a distributed, grassroots organizing model. Th!rdAct's mission aligns closely with the goal of this book—although the target audience—"experienced Americans"—is broader than "boomers" and the Th!rdAct's concerns include protecting democracy as well as climate. The organization's homepage says it best:

> *Th!rdAct is people over the age of 60—"experienced Americans"—determined to change the world for the better... We muster political and economic power to move Washington and Wall Street in the name of a fairer, more sustainable society and planet. We back up the great work of younger people, and we make good trouble of our own...*

McKibben articulated the reasons for older Americans to act on climate long before he founded Th!rdAct. "It would be entirely fitting if the angry troublemakers came from the ranks of those of us who are older," he wrote in a 2006 essay titled "A Last Best Chance for Baby Boomers."

"For one thing," he continued, "we're the ones who caused the problem. If someone has to sit down on the tracks of the coal train and get arrested it should be the grandparents who have been pouring carbon into the atmosphere for half a century. But there's something else, too: the greatest moments in the lives of the baby boomers were precisely the times when they raised their voices, when they declared their selfless devotion to peace or civil rights."

I no longer remember when I first came across that essay but remember that the words made a deep impression on me, articulating something that I already believed deeply. The enthusiastic response to McKibben's launch of Th!rdAct suggests that I am not alone.

Like other organizations described in this chapter, Th!rdAct provides tools and connections to support new activists, encouraging volunteers to organize working groups based on geography or affinities, such as artists, educators, and lawyers.

Soon after I signed up, I volunteered as a co-organizer of Th!rdAct's Northern Virginia hub and was then invited to serve on the state-wide coordinating committee. Together, the eight of us on the committee are learning how to be more effective organizers, while planning and carrying out activities to advance Th!rdAct campaigns. In 2022 this included supporting climate legislation that eventually passed as the Inflation Reduction Act, turning out climate voters in the mid-term elections, and opposing the "Dirty Deal"—permitting reform legislation that would grease the skids for new fossil fuel projects.

Th!rdAct can already claim a notable success. In February 2022, McKibben suggested that President Biden use the 1950 Defense Production Act (DPA) to spur production of heat pumps to ease energy shortages resulting from Russia's invasion of Ukraine, while accelerating the energy transition.[41] All-electric and highly efficient, heat pumps replace gas furnaces and double as air conditioners. Thousands of letters from Th!rdAct members helped bring the idea to the attention of the president, who in June invoked the DPA to boost production of five key clean energy technologies including heat pumps, solar, insulation and transformers and other electric grid components.[42]

"Ideas do absolutely no good unless lots of people get behind them and make noise," McKibben wrote to Th!rdAct members. "That's what movements are—people making impossible things possible. I'm immensely grateful to everyone who wrote and pestered their politicians about this; it worked."

In early 2023 Th!rdAct was pressuring the four big "climate-bad" banks to stop financing fossil fuels, not only by encouraging older Americans to move their money, as described in the previous chapter, but by organizing demonstrations and rallies in front of the banks' retail outlets.

One Th!rdAct volunteer advancing this effort is Denise Duarte, a Las Vegas-based multi-disciplinary artist. Duarte's artwork has often explored LGBTQ themes, such as a recent solo show that examined sexuality and gender identity using the botanical world as metaphor. Since volunteering for Th!rdAct she has served on the arts working group, helping to design protest art, while organizing a new Th!rdAct working group focused on Las Vegas and nearby areas, home to three-out-of-four Nevada voters.

What motivated her to become a Th!rdAct volunteer? "We put solar panels on our house. We got a hybrid car. We had done most of what we could do to reduce our carbon footprint," she recalled. "Seeing what was happening, I got more and more concerned. When my spouse died in April, I was left adrift. I felt that I needed to do more." When she received an email about Th!rdAct, she thought: "This is the missing piece. This is what I need to do." Having devoted her career to creating art dedicated to social justice, she realized "it doesn't matter what we do on any social justice issue if humanity doesn't survive."

Many of the Th!rdAct volunteers I have met have reached similar conclusions, regardless of their background. They include a mental health professional who retired early to become a full-time climate activist, a retired small business owner, an adult literacy expert who previously taught in prisons, and an engineer.

Whether or not you decide to make it your primary affiliation, I hope that everybody who reads this book who is not already signed up on Th!rdAct's website will do so. It's a great way to meet other climate boomers and find out what they are doing to save the planet. An added bonus: listening to McKibben on the monthly Th!rdAct all-in calls and reading his newsletters will keep you motivated and inspired to do more.

Build Consensus with Citizens' Climate Lobby

Citizens' Climate Lobby (CCL) is one of the largest grassroots climate advocacy organizations in the United States, with more than 60 paid employees, some 500 chapters, and 200,000 supporters. Maybe you never heard of it? That's because CCL is resolutely non-partisan and works behind the scenes, quietly engaging with members of Congress regardless of party.

Since its creation in 2007, CCL has built political support for pricing carbon pollution, specifically a national, revenue-neutral carbon fee and dividend. Economists have long argued that pricing carbon emissions creates powerful incentives for rapid decarbonization, for example, by accelerating the rollout and uptake of new technologies. CCL proposes that fees collected from fossil fuel companies be distributed on an equal per capita basis to every American. These "carbon dividends" would protect low and middle-income Americans from higher energy prices during the transition.

Progressive climate groups tend to distrust carbon pricing and some actively oppose it, concerned that it will exacerbate environmental injustices by strengthening incentives for polluting industries to locate in low-income communities, especially those with more people of color. The hostility intensified in 2021 after a senior Exxon lobbyist was caught on camera saying that oil companies pretend to support the policy, because they believe it will never pass.

While I share these concerns, I agree with most economists who have studied the issue that carbon pricing could greatly accelerate the energy transition around the world, through a mechanism known as Border Carbon Adjustment (BCA). In simple terms, a BCA would tax carbon-intensive imports from countries that don't impose a carbon fee. To avoid paying these border taxes, exporting countries would impose their own fees, setting off a virtuous cycle in which charging for carbon pollution rapidly becomes the new global norm—and emissions plummet.

In 2014, I imagined how this might play out in a piece of political science fiction titled *The Sudden Rise of Carbon Taxes (2010–2030)*.[43] In this essay, a Chinese co-author and I looked back from 2030 and described how the United States and China each imposed carbon pollution fees and corresponding BCAs, prompting other countries to follow. With high and rising prices on carbon pollution around the world, emissions drop and new technologies for capturing carbon from the atmosphere are deployed quickly, averting climate disaster.

That clearly hasn't happened—yet. Moreover, CCL's efforts to build a consensus for carbon pricing, by nurturing a bi-partisan House Climate Solutions Caucus, with an equal number of Democratic and Republican members, suffered big setbacks when the moderate Republican members lost their seats—those in purple districts to Democrats, those in red districts to pro-Trump climate deniers. The coalition collapsed and a Conservative Climate Caucus that is so far mostly a greenwashing exercise arose in its place.

But carbon pricing and BCAs are far from dead. In December 2022, the European Union approved landmark legislation that would impose fees on carbon-intensive imports such as iron and steel, cement, aluminum, fertilizers, electricity, and hydrogen. The law will begin to take effect in late 2023, when the EU will collect data on the "embedded carbon"—the carbon emitted during production—of the targeted imports. Importers will begin to pay the fees in 2026.[44]

Long before then, EU trade partners, including the United States, will need to decide what to do about the impending tariffs—including whether to move ahead with their own version of a carbon price. That dynamic, combined with a possible GOP retreat from Trump-inspired extremism following the party's poor performance in the 2022 midterms, could put carbon pollution fees back on the political agenda. If that happens, CCL's national network of trained citizen lobbyists will be ready to help it gain traction.

In the meantime, CCL has announced that it is broadening its policy goals, adding "healthy forests, building electrification and efficiency, and clean and fast permitting" to their list, while keeping carbon-price-and-dividend as the primary focus. "Clean and fast permitting" is a reference to the contentious debate about how to streamline permitting for renewable energy projects without opening a new door for fossil fuels or running roughshod over environmental protections. Searching for a bipartisan consensus on tough issues such as these is where CCL excels.

CCL executive director Madeleine Para, who as a volunteer in 2011 founded a CCL chapter in Madison, Wisconsin, recalled her experience in an online address to the organization's annual conference in December 2022. "I remember how I felt before finding CCL," she said. "I was frustrated because I didn't feel like protesting was enough, and I didn't want to treat people who disagreed with me like they were my enemy."

"When I heard that I could become a citizen lobbyist to my member of Congress and work by building relationships, I jumped in and learned everything I could... I stopped feeling so alone, and I started feeling like being part of a big and growing team." Para explained that CCL would continue to champion carbon pricing, while also pursuing the four new policy initiatives, which are likely to be on the legislative agenda, providing more ways for CCL citizen lobbyists "to work with your community and with Congress."

Although CCL welcomes participants of all ages—and has made a big effort to recruit younger members as well as people of color, boomers, mostly white, have been the backbone of the volunteer corps. Many are attracted by the non-confrontational approach that Para describes.

For example, Marc Peterson, a volunteer in Utah, had a long and successful career in the energy industry, leading a corporate team that sold heavy generation equipment, first for fossil fuels and later for renewables, mostly wind, and eventually batteries. He enjoys using his deep knowledge of the energy industry to explain why rapid decarbonization is necessary—and potentially profitable.

For the first three years after his retirement in 2016, however, he says he "did nothing." His corporate job had been all-consuming and he wanted to take some time off. "But after a while I felt like I wasn't adding value to the world. To maintain my sanity, I needed to feel that I was contributing to society." A friend introduced him to CCL and "the more I learned, the more I felt comfortable."

He estimates he spends a couple of hours a day volunteering for CCL, as co-lead of the Salt Lake City chapter and helping with regional organizing. He and fellow volunteers write letters to the editor, "table" (i.e., pass out fliers and answer questions at public events), and make presentations "everywhere we can."

Peterson is in high demand because of his deep knowledge of the energy industry—he eagerly cites new data to demonstrate that renewables are indeed cheaper than fossil fuels. Politicians in Utah are increasingly open to these arguments, he says, because climate impacts, such as less snow for the state's famed ski resorts and rapidly falling water levels in the Great Salt Lake, are becoming ever more obvious.

"Having a background in energy and understanding the science, I have a moral obligation to act," he says. "Just because we are boomers doesn't mean we can't add value to the world."

Join a Grassroots Group—Or Start One!

Th!rdAct and CCL mostly aim to achieve national-level policy change. But many of the most immediate and extreme impacts of the fossil fuel industry happen at the local level. Frontline communities are those that are most harmed by climate impacts and fossil fuel pollution. Mostly low-income and communities of color, they face a range of environmental and health hazards due to toxic chemicals released during the extraction, production, and use of fossil fuels.

Chances are that there is a community like this in your state—indeed, you may live in or near one. If you do, finding and joining in the work of a grassroots group opposing fossil fuel projects can be a powerful way to fight for environmental justice and to slow climate change. Live in a front-line community where no such organization exists? Perhaps you can start one!

That's what Sharon Lavigne did. Born in Louisiana's St. James Parish in 1950, Lavigne recalls a childhood when her family ate fish from the river and vegetables from their garden, and people rarely got sick. Starting in the 1960s, petrochemical plants began opening in the area until there were some 200 along an 80-mile stretch of the Mississippi River that came to be known as "Cancer Alley." According to the EPA, cancer risk for the area's Black population is 50 times the national average.

"When I realized those people were dying, I couldn't understand. I thought maybe the world was coming to an end. And no one was doing anything about it," Lavigne recalls in a documentary about her work. "Neighbors on both sides of me had cancer. They passed away. Friends, loved ones... Just about every household you would speak to in St. James had someone that was lost to cancer."[45]

In 2018, the governor announced plans for a new plastics plant, the state's biggest, just three miles from her house that would disrupt a cemetery that is the burial site of enslaved ancestors. When the local government quickly granted permits for the $1.25 billion plant, Lavigne, a Catholic, decided to fight back. She organized RISE St. James, a faith-based advocacy group to block it.

The group started in October with a meeting in her living room. Ten neighbors showed up and her daughter took notes. Early members recruited others to join, educated others in the community, and organized peaceful protests. One year later, in response to an intense community campaign, the company canceled the project.

Lavigne knows all too well the connection between the industrial plants she is fighting and the emissions heating the planet. In August 2021, Hurricane Ida, one of the most powerful storms in Louisiana's history, tore the roof off her home. "While big banks are making the big bucks financing fossil fuels, our communities in Cancer Alley are suffering financially in a big way from climate change," she later said.

Just two months before, Lavigne received the Goldman Environmental Prize, also known as the Green Nobel, one of six awarded around the world each year to grassroots activists. In 2022 the University of Notre Dame awarded her the Letare Medal, the most prestigious award for American Catholics. She continues to fight the expansion of the petrochemical industry in her parish and surrounding areas, hosting a "Chemical of the Month" event for her neighbors to learn more about the toxins in their environment.

Engage with Your Alma Mater

For those of us who attended college or university, supporting a student-led divestiture campaign at your alma mater is a great way to become a climate activist. Chances are that your school alumni office regularly invites you to stay in touch. They do this for a reason: alumni gifts are a key source of funding. Whether or not you give to your alma mater, the school's interest in you as a donor or potential donor gives you special leverage.

One thing the alumni office won't do is invite you to become involved in a conversation about fossil fuel divestiture. Alumni offices exist to raise money and those who work there are rewarded for maximizing gifts, not for thinking about the impact of the school's investments. Yet fossil fuel divestiture campaigns are underway at thousands of institutions of higher learning across the country. All welcome support from boomer alumni.

Our generation invented modern divestiture as a tool to tackle injustice. In the late 1960s, American students began urging their colleges and universities to divest from companies that did business with the apartheid regime in South Africa. In 1977, Hampshire College, a private liberal arts college in Amherst, Massachusetts, became the first U.S. institution of higher learning to divest from companies doing business with South Africa, selling a modest $39,000 in stocks it held in four companies.

As resistance to apartheid within South Africa gained momentum, more than 200 U.S. institutions of higher learning followed suit. For example, in 1986 the University of California, responding to years of agitation by students, authorized the withdrawal of $3 billion of investments from the apartheid state. Nelson Mandela later said that the UC's move helped end white-minority rule in South Africa.

More than 30 years later, tiny Hampshire College once again led the way, becoming the first institution of higher learning to announce that its assets were fossil-free. Divesting from South Africa had let the college adopt an early form of what we now call ESG investing. Jonathan Lash, president of Hampshire College at the time, wrote that when Bill McKibben visited the school in 2011 and asked students to join the 350.org divestiture campaign, the school's modest endowment was already fossil-free.

"We checked and found our policy had already led us away from these investments," he wrote. "Our students celebrated this finding, and as a result, we were widely credited as the first college to be divested from fossil fuels."[46]

Since then, the movement has spread across the country—and inspired fossil fuel divestiture efforts in other sectors, including pension funds and religious institutions. By late 2021, 220 educational institutions had committed to divest $6.2 trillion in endowments and other financial assets from fossil fuels.[47] (Stand.earth, which encourages divestiture, reports total assets that commit to divest, rather than the value of the stocks sold as a result, because the latter is too hard to track.)

The movement claimed an important victory in 2021, when Harvard, the richest university in the world, yielded to pressure from students and alumni and announced that it would phase out fossil fuel holdings in its $42 billion endowment. Harvard's president, Lawrence Bacow, said that the university no longer directly owned any fossil fuel investments and "does not intend to make such investments in the future."

A messy debate ensued about whether the university had really shed its fossil fuel investments. Critics noted that 60% of Harvard's endowment is entrusted to private equity and hedge funds that may have different investment strategies. Others, including fossil fuel critic and Harvard professor Naomi Oreskes, said that the announcement still mattered, even if it amounted to less than full divestiture. "Harvard's divestment is a signal to other investors that as the planet burns, finance must not stand with the arsonists," she wrote in *The New York Times* op-ed.[48]

Danielle Strasburger, co-founder of Harvard Forward, an alumni-led group that supported the student-led divestiture movement, said that having older alumni involved was important to success. "A lot of our biggest supporters and stellar volunteers were older alumni," she told me. "It was a really great cross-generational endeavor, especially getting to learn from older alumni about their previous campus activism around South Africa divestment and the Vietnam War."

"Often older alumni have resources, skills, experiences, connections, and influence that students and young alumni haven't had time to build up yet, so they're a key piece of the movement," she added. "We felt like the university often tried to wait out or downplay student activism. Building a coalition that included students, faculty, and alumni across generations helped show that it was a priority for the whole community and should be a priority for the university."

Harvard activists are now taking aim at a more insidious relationship between their university and the fossil fuel industry: influence buying. "Universities like Harvard are entangled with the industry in more ways than their endowments," according to a student report released in 2022. "Even as they move to make their investments fossil free, they are allowing fossil fuel companies to fund research and programming, including around key issues of climate science and policy."[49]

Examples included extensive fossil fuel funding at the Harvard Kennedy School, Shell and BP funding the Harvard Environmental Economics Program, and ExxonMobil and Chevron funding the university's Corporate Responsibility Initiative. The report described how professors who receive large grants from fossil fuel companies discourage student critiques of the industry.

Student activists are seeking alumni support for two policies to limit such blatant conflicts of interest. First, that the university disclose data about fossil fuel grants, gifts, donations and funding for research. And second, that it stop accepting fossil fuel money for research related to energy policy, climate change, and the environment. Harvard Forward has created a website, DivestHarvardAlumni.com, to appeal for alumni support.

Meanwhile, Harvard's divestiture announcement has energized similar campaigns at other universities. In early 2022, student climate activists at Yale, Stanford, Princeton, MIT and Vanderbilt filed legal complaints to compel their universities to divest. Like Harvard Forward, these and other student-led efforts all welcome alumni support.

Don't know if your school has already divested? The Fossil Fuel Divestment database run by Stand.earth and 350.org tracks more than 1,500 institutions with total assets of more than $40 trillion that have committed to divesting from fossil fuels. The database includes pension funds, religious institutions and other types of organizations in addition to educational institutions. Many listings include a link to the relevant announcement, which can lead you to additional information. If your institution isn't on the list, it probably hasn't announced plans to divest.

To find campaigns currently underway, *GoFossilFree.org* provides a database of some 387 campaigns in the United States and elsewhere. Or you can do an internet search for your university and fossil fuel divestment. If a student group is already advocating for divestiture, reach out to ask how you can help. If you find nothing, try writing to elected student government officers, asking them to put you in touch with a student-led climate group. Don't forget to engage with your former classmates—you are likely in touch with several on Facebook—sharing your concerns and asking them to join your efforts.

Perhaps you will be the one to launch an alumni campaign for your alma mater to divest from fossil fuels and limit the conflicts of interest that arise when fossil fuel companies fund teaching and research.

Tell Your Pension Fund to Clean Up Its Act

If you are lucky enough to have money in a pension fund, chances are that the fund is invested in fossil fuels. According to the Climate Safe Pensions Network, with more than $46 trillion in assets worldwide, pension funds are among the largest institutional investors in fossil fuels, providing the industry with the financial capital and social license it needs to operate. In the U.S. alone, pension funds are worth $4 trillion—and almost all have fossil fuel holdings. "As a result, our tax dollars and retirement funds are being used to support the powerful coal, oil and gas companies that are causing the climate crisis, polluting our communities and violating Indigenous rights," the group notes. "Until pension funds divest from fossil fuels, they are complicit in the climate crisis."[50]

Don't underestimate your potential power as a pension beneficiary. Your status gives you standing to demand that the fund managers disclose their holdings, divest from fossil fuels, and reinvest in renewable energy and other climate-friendly businesses. After all, it's your money!

This is true whether you are part of a traditional "defined benefit" pension that pays a set amount based on salary and years of service, or the more common "defined contribution" tax-protected retirement savings plans, such as 401(k)s and Roth IRAs, that pay based on how much you have contributed and the fund's performance.

Either way, you can tell those who manage the fund you do not want your money invested in businesses that destroy the future. You also can object that the fund's fossil fuel holdings are putting your retirement savings at risk, because fossil fuel companies are a bad long-term bet. Because proven fossil reserves cannot be burned without cooking the planet, divesting from fossil fuels protects your retirement savings from potential losses associated with these stranded assets, while opening opportunities to invest in sectors with better long-term potential, including clean energy, climate change adaptation, and carbon sequestration.

The first step in applying your climate activism to pension fund divestment is to determine where your retirement savings are held. You likely know this already. Many of the big funds were initially set up to provide benefit-defined pensions for public-sector employees, such as teachers, fire fighters, and other government employees. As employers shifted increasingly to contribution-defined plans, the pension funds began offering these tax-protected retirement savings plans as well.

Once you know the name of your pension fund, check the list of divestiture campaigns maintained by Climate Safe Pensions. Most are based on geography—California, Colorado, Maine and New Jersey are among the states with active campaigns.

Others are focused on specific funds. For example, the TIAA-Divest! Campaign is working to stop the retirement giant TIAA (Teacher Insurance Annuity Association) from financing climate destruction. TIAA's clients are mostly employees in universities and other not-for-profit organizations, and it has branded itself as a socially responsible investor. According to TIAA Divest, it manages $1.3 trillion in assets and is one of the largest owners of farmland and timberland, while investing billions of dollars in oil, coal and fracked gas.

Each campaign is run independently and offers a variety of ways to get involved. These include sign-on letters, online briefings and trainings, and appeals to join in public action. If you find a campaign underway for your pension fund, you have found kindred spirits! Sign their petition, attend their training sessions, and ask how you can help. Recruiting others is a great way to use your existing networks. Whether you are an early boomer who retired years ago or a late boomer who is still working, reach out to friends and colleagues who are beneficiaries of the same pension fund. Explain why this matters to you and invite them to join you in a specific activity, such as signing a petition, joining a training, or attending a public protest.

What if you live in a state, or belong to a pension fund, that doesn't yet have a campaign? You can be the spark that lights the prairie fire. Fossil Free Pensions offers tools and guidance to start your own campaign, including a campaign-plan template, social media templates, guidelines for purchasing ads on social media platforms, and design files to give your materials a professional look. Don't be afraid to introduce yourself and ask for guidance using the simple form on the website.

If you worry that such campaigns are mere tilting at (or in favor of!) windmills, know that change is coming fast. In December 2021, New York City Comptroller Scott Stringer announced that three of the city's five biggest pension funds had divested about $3 billion in fossil fuel company holdings. The divestment, one of the largest in the world, followed previous commitments by some of the funds to divest from coal and tar sands oil. The decision shows that "environmental and fiscal responsibility go hand-in-hand," Stringer said in the press release announcing the move. "New York City is leading the way toward a clean, green and sustainable economy, and the impacts of the actions we are announcing today will be felt for generations to come."

Use Your Professional Ties: Medicine, Law, Advertising

I round out this chapter by noting three opportunities to use your professional connections to fight climate change.

First, healthcare: Doctors, nurses and other healthcare professionals are among the most trusted professionals in America, and climate change is a large and growing threat to human health. Medical professionals are therefore uniquely placed to ensure that the health risks of climate change and the benefits of climate solutions, especially clean energy, are clearly understood.

The Medical Society Consortium on Climate and Health brings together associations representing more than 600,000 health professionals to deliver three messages:

- Climate change is harming Americans and these harms will increase unless we act;
- The way to slow or stop these harms is to decrease the use of fossil fuels and increase energy efficiency and use of clean energy sources; and
- These changes in energy choices will improve the quality of our air and water and bring immediate health benefits.

The consortium trains doctors and other healthcare providers to become advocates in their communities by speaking to policymakers, press and community groups. It also issues statements on behalf of the several dozen medical associations. On Earth Day in 2022, for example, the consortium released a letter to the Biden administration along with 58 other health organizations urging more investment in clean energy—and less in fossil fuels—to achieve energy independence.

"The early leaders of our group were baby boomers who were veterans of advocacy work against the tobacco industry, or the Vietnam War, or apartheid, or in favor of greater access to health coverage or opportunities for women and minorities," consortium founder and former executive director Mona Sarfaty told me in an interview.

They had seen how the tobacco industry used "doubt-sowing tactics" to fight off regulatory attempts and saw the fossil fuel companies doing the same thing with global warming, she said. These founders surveyed doctors and discovered that they were already seeing climate-related health impacts in their patients, such as exposure to heat, flooding, air pollution from fires, and more disease from insects.

"The doctors wanted to be involved in speaking publicly about climate policy," she said. "They believed doctors should inform the public and their own patients about these risks. This knowledge helped us form a coalition of medical societies."

Today the consortium serves a multi-generational group of health professionals, with many young professionals increasingly active and concerned, she said. Boomers are still very welcome and can get involved through their own medical professional society or by visiting the consortium website, she said.

The second professional stream is the law. Fossil fuel interests rely on top law firms to avoid being held liable for their destruction of the planet. Law Students for Climate Accountability (LSCA), a network of students in more than 50 law schools, aims to "stigmatize and ultimately eliminate the legal industry's complicity in perpetuating climate change." It asks law students to pledge not to work for firms that do this dirty work or, if circumstances prevent them from doing so, to push the firm to stop representing fossil fuel interests.

LSCA publishes an annual scorecard grading the country's top 100 law firms based upon their litigation, transactional and lobbying work on behalf of fossil fuel companies. In 2022, only two firms earned an "A": Cooly, and Shulte Roth & Zabel. Firms earned an "F" if they handled eight or more cases exacerbating climate change, supported more than $20 billion in fossil fuel transactions, or received more than $2 million for fossil fuel lobbying.

The country's top two law firms by revenue, Kirkland & Ellis and Latham & Watkins, each earned an "F." The third-ranked firm by revenue, DLA Piper, earned a "D." In all, 38 firms performed the extraordinary amount of fossil fuel work necessary to fail, two more than in 2021; 36 firms received a "D."

The scorecard argues against the notion that companies are entitled to representation. "There is wide agreement among legal ethics scholars that unless indigent criminal defendants or court appointments are involved, no lawyer has an obligation to represent any particular client," it states. When Russia invaded Ukraine, it notes, most international law firms closed or suspended their Moscow offices and several dropped Russian clients.

LSCA national director Haley Czarnek told me that boomer lawyers, retired or not, can do a lot to advance the group's efforts. "Older lawyers have the knowledge and comfort to make choices," she says. If you are still practicing, "quit doing fossil fuel work as quickly as possible," she says. Boomers can also help by mentoring law students and younger lawyers facing difficult career choices related to fossil fuel representation, she adds. Finally, they can talk to their peers.

"They know what matters to those people. Maybe it's their kids, maybe it's skiing every year, maybe it's the reputation of the firm," she says. Use this knowledge—and the rankings in the LSCA report card—to nudge firms away from fossil fuel work, she advises.

The third professional stream is advertising and public relations. Fossil companies use advertising and PR agencies to gain the tacit approval for their activities known as "social license." Although fossil fuel companies are the world's biggest climate polluters and actively oppose policies to avert catastrophe, their advertising campaigns falsely assert just the opposite—that they are clean, green, and actively working for climate solutions.

Clean Creatives, a network of people and companies working in advertising and public relations, invites individuals—known in the industry as "creatives"—and the agencies that employ them to refuse such work. The pledge is simple: "We will decline any future contracts with fossil fuel companies, trade associations, or front groups." By the end of 2022, some 1,300 creatives and more than 450 agencies had taken the pledge.

Clients—companies that buy advertising and public relations services—also have a role to play, refusing to hire agencies that work for fossil fuel interests. "When agencies help Big Oil with misleading greenwash campaigns, they make it harder for consumers to identify and support brands making genuine investments in social good," the Clean Creatives website explains. "Clean agencies can help you build your brand, without contributing to fossil fuel pollution."

Boomers who worked in advertising and public relations industries can support Clean Creatives by spreading the word and encouraging the firms they worked with to take the pledge.

Ready, Set, Go!

The chapter you have just completed is the core of the boomer's guide to climate action. I began by suggesting you think about what you bring to the climate movement, your strengths and connections. I followed with guidance on how to use national organizations to find local connections, listing several and highlighting two, Th!rdAct and Citizens' Climate Lobby.

Along the way I shared conversations with people who have done what I'm asking you to do: become a climate activist. They come from many walks of life and contribute in a variety of ways. The chapter concluded with four channels for you to exert your influence: joining or starting a grassroots group in a frontline community, engaging with your alma mater, pushing your pension fund to clean up its act, and using your professional connections in areas such as healthcare, law and advertising.

I hope you have been marking up the book as you go, highlighting the things that resonate with you. I suggest you take a moment now to jot down just three things from this chapter that you will follow up on. You can write them on this page or in a notebook that you have begun to track your climate action journey.

The next chapter tells why religion can be an important source of power for the climate movement, and explains how you can tap into this, whether or not you are currently active in a faith-based community.

Chapter 5: How Can I Work with Others?
Action Checklist

- Assess your strengths and networks
- Use national organizations to find local contacts
- Join Th!rdAct or another group pushing for rapid change
- Support a frontline grassroots group—or start one
- Engage with your alma mater
- Tell your pension fund to clean up its act
- Leverage your professional connections

Chapter 6

What's faith got to do with it?

All religious traditions require we protect the planet and care for the poor. Top faith leaders are urging action. Faith-based climate advocacy can be a source of power: when religious groups demand action, politicians listen.

Chapter 6

What's faith got to do with it?

So, a priest, a rabbi and an imam walk into a bar. Bartender goes: "What is this, a joke?" Seriously, as old jokes like this suggest, when leaders from different faiths act together, people pay attention. If a priest, a rabbi and an imam—or other representatives of multiple faiths—walk into a Congressional office to talk about climate change, or jointly sign a resolution, or march together in a climate protest, political leaders and others pay attention.

Why? Unlike nearly everybody else trying to shape public policy, faith-based groups, including interfaith groups, are understood to be acting out of moral conviction grounded in religious teaching. This means that religious voices can be a powerful force for advancing climate action. You can tap into this force whether you are already an active participant in a religious community, only very loosely affiliated with a faith-based tradition, or perhaps just beginning to think about launching a late-in-life spiritual quest.

Faith-based action offers other benefits besides a path to power. As we shift into retirement, many of us lose ties with our workplaces and former colleagues. We find ourselves searching for community. For some of us this can mean reconnecting with the religious teachings, rituals, and communities of our childhoods. For others it may entail a spiritual search that leads to a new community and set of traditions. Boomers who relocate during retirement often find that joining a church, mosque, synagogue or other religious organization provides ready entry to a welcoming community. Choosing a community with a commitment to climate action can align your values with your spiritual or religious practice or quest.

Finding a Path to Faith-Based Advocacy

Joelle Novey is a grassroots organizer who serves as the director of the DC, Maryland and Northern Virginia hub of Interfaith Power and Light, a national organization that engages faith communities in environmental stewardship and climate action advocacy. She suggests two steps for boomers who want to explore faith-based climate advocacy. First, "seek out a grassroots group that meets regularly in your area and will call you to action," whether within your own faith tradition or interfaith. Second, "seek out the language and theology that is authentic to your faith tradition—all faiths have it—and use that to ask your faith leaders to do more."

If you do this, you will have plenty of company. A 2021 *Climate in the American Mind* survey found that most religious voters favored legislation to eliminate fossil fuel emissions by 2050, often by overwhelming majorities. This included 88% of Black Protestants; 76% of non-Christian religious groups; 61% of white Catholics, 53% of white ecumenical Protestants; and 50% of white evangelical Protestants.

Although Native Americans don't appear in such surveys, Indigenous people, especially elders, have been prominent in the fight to protect the Earth, often invoking their sacred teachings and traditions. Native leadership in efforts to block fossil fuel pipelines such as Keystone XL, Dakota Access Pipeline, Line 3 and others have helped to inspire, educate and mobilize thousands of Native and non-Native people to join the fight.

Although Native American religious beliefs differ widely, and European settlers and the U.S. government banned and violently suppressed Indigenous religious practices for centuries, core concepts have persisted and some have made their way into the majority culture. One is the belief that the entire universe is alive and therefore sacred. Another, based on an ancient Haudenosaunee (Iroquois) philosophy, requires humans to consider the impact of our decisions on people seven generations into the future.

In fighting for their lands and sovereignty, Indigenous peoples are fighting to save the planet. Although they comprise less than 5% of the world population, Indigenous peoples protect 80% of the Earth's biodiversity in the forests, deserts, grasslands, and marine environments in which they have lived for centuries.[51]

One well-known Indigenous elder is Winona LaDuke, a rural development economist, environmentalist, writer and activist who has authored and co-authored more than a dozen books. An Anishinaabekwe (Ojibwe) enrolled member of the Mississippi Band Anishinaabeg, she is the co-executive director of Honor the Earth, which she founded in 1993 to increase awareness and raise money for Native-led environmental groups.

I heard her speak at the interfaith camp of the Treaty People's Gathering in northern Minnesota in June 2021. It was an alarmingly hot day. We sat beneath pines beside a lake on rough-hewn benches in a campfire-style circle. A middle boomer with a commanding, charismatic presence, LaDuke held our group of about 100 mostly non-Native climate activists spellbound.

In one of her earliest books, *Recovering the Sacred: The Power of Naming and Claiming* (2005), LaDuke writes that because religious freedom is a fundamental part of the U.S. Bill of Rights one would think that religious freedom for Native people is protected. "That is so, as long as your religious practice does not involve access to a sacred site coveted by others," she wrote. The book recounts scores of instances where Native people's rights to sacred lands have been blocked or otherwise violated. It ends with a plea for action.

"The fossil fuel century has been incredibly destructive to the ecological structures—the air, earth, water, and plant and animal life—that keep planet Earth habitable for humans," she wrote. "Whether human populations will continue to flourish 100 years from now will depend on the choices we make today." Fifteen years after that book was published, her plea is truer—and more urgent—than ever.

Another group that has been influential in the climate movement beyond what their numbers would suggest is the Quakers, formally known as the Religious Society of Friends, who have been early movers in the fossil fuel divestiture movement. The Friends Committee on National Legislation, a non-partisan Quaker group that lobbies Congress, has actively supported the Civil Rights Movement and Native American rights, and opposed nuclear weapons and the Vietnam War. Today the group also lobbies for climate action.

American Buddhists have expressed extreme concern about climate change but do not currently have an organized response. A 2015 declaration *The Time to Act Is Now* signed by prominent Buddhist leaders, including the Dalai Lama, sparked the launch of a group called the Buddhist Climate Action Network. A 2019 posting on the group's Facebook page stated that "the Climate Breakdown is even more serious and advanced than we knew." Given this urgency, the post said, American Buddhists should join other groups, like Extinction Rebellion, rather than organizing their own activities.

At the local level, many congregations focus on reducing the climate and environmental footprint of their buildings and grounds, through measures like recycling and rooftop solar. While this approach is understandable, too many congregations get stuck there and fail to mobilize the potential power of faith-based advocacy to demand systemic change. Given how high the stakes are, I believe that religious institutions could be doing much, much more.

This creates opportunities for climate boomers. Older people are the majority of active participants in religious organizations and provide the cornerstone of financial support as well. Clergy and other religious leaders are often personally concerned about the climate emergency but they pick their priorities based on the guidance from their leadership and the expressed needs of their followers. Many are willing to be nudged to do more. You can make a difference by providing that nudge!

In the pages that follow I offer an overview of how several faiths in the United States are responding to the climate emergency, starting with the larger groups.[52] There are two ways to read this chapter. If you identify with a faith tradition, skipping to the section that matches your background is reasonable. But I hope that many readers—regardless of where they find themselves on the spectrum of belief and non-belief—will want to understand the climate-related teachings and actions of the faith traditions I discuss below. After all, saving the planet will require that people of all faiths and none join in a common effort.

Catholics

Issued in May 2015, *Laudato si'* (Praise Be to You), the second encyclical (open letter) of Pope Francis, is arguably the most important statement on climate change ever issued by a major religious leader. Its influence reached well beyond the Catholic Church and it continues to inspire Catholic climate action today. In the encyclical, which has the sub-title "On Care for Our Common Home," Pope Frances calls all people of the world to take "swift and unified global action" to address climate change. He also critiques consumerism and irresponsible development, and laments environmental degradation.

"Never have we so hurt and mistreated our common home as we have in the last two hundred years," Pope Francis wrote. He calls for highly polluting fossil fuels "especially coal, but also oil and, to a lesser degree, gas" to be "progressively replaced without delay." The encyclical frames climate change as a social justice issue, stating that developed nations are morally obligated to help developing nations combat the crisis. Linking poverty and the environment, Pope Francis insists that the world must "hear both the cry of the earth and the cry of the poor."

Unlike most encyclicals, *Laudato si'* found an audience well beyond the Catholic Church. "On a sprawling, multicultural, fractious planet, no person can be heard by everyone. But Pope Francis comes closer than anyone else," wrote Bill McKibben, who identifies as a Protestant. The pope's letter, he added, was "nothing less than a sweeping, radical, and highly persuasive critique of how we inhabit this planet—an ecological critique, yes, but also a moral, social, economic, and spiritual commentary."

For Dan Misleh, founder of the Catholic Climate Covenant, a lay-led organization that helps U.S. Catholics respond to the Church's call to care for creation and care for the poor, *Laudato si'* was a dream come true. "I used to literally dream that a pope would write an encyclical on the environment," Misleh told me. Although concern about the climate change and the environment were not new to the U.S. Catholic Church, the power of the encyclical's language and the prestige of the pope brought new energy into the Catholic climate movement, he said.

Boomers who have ties to the Catholic Church have a critical role in helping the church to up its game on climate, said Misleh. "Boomers have gone through their careers, they have wisdom, money, and a certain level of clout within their Catholic community," he said. In every parish, "they are the age group that is most active in the Church."

There are plenty of ways to engage. The Vatican maintains a *Laudato si' Action Program* website that includes resources ranging from Vatican updates to tools for completing a "self-assessment" and uploading "reflections" and a "personal action plan." The Catholic Climate Covenant offers GodsPlanet —tagline: "We are all part of God's plan(et)"—a website that celebrates and supports American Catholics acting on climate, from organizing LED lightbulb refits and installing solar on churches and schools to advocating with Congress for passage of climate legislation.

"Our goal as an organization," said Misleh, "is to make it as easy as possible for Catholic leaders and other Catholics to get involved."

In 2022, the Catholic Climate Covenant organized volunteers to meet with their senators to urge support of the climate-focused Build Back Better legislation that became the Inflation Reduction Act. "Senators and staffers have been receptive to hearing about Catholic Church teachings on climate change and care for creation," according to an account of the meetings released by the organization. "They have responded well to the faith-based aspects of the meetings, which include an opening prayer calling the Holy Spirit into the group's dialogue."

In one such meeting, volunteers met with Indiana Sen. Mike Braun, a conservative Republican who has said his party should take climate change more seriously but also backed President Trump's withdrawal from the Paris Agreement. It was the largest virtual meeting that Sen. Braun had participated in, with people from across the state, including young people, professors, priests and two Catholic bishops.

Misleh is quick to acknowledge that the church's response to climate change falls short of what is needed. Part of the problem, he says, is that there are many other issues that the church is concerned about: immigration, abortion, the fallout from the sex abuse scandals. Catholic boomers can help the Church to do more on climate, he says, through a "top down, bottom up" dynamic where the "people in the pews" cite the Pope's encyclical and encourage Catholic leaders to publicly back demands for more ambitious climate action.

Catholic voices for climate action have the potential to be extremely influential. More than one out of five Americans identifies as Catholic and, according to one study, nearly half of all Americans say that they have at least some connection to Catholicism. Catholics span the ideological spectrum, from highly conservative to fiercely liberal and unlike evangelicals and some other Protestants, who tend to have racially distinct denominations, all U.S. Catholics belong to the same institution, regardless of race and ethnicity. These unifying traits mean that Catholic climate advocacy can help bridge partisan and other social divides.

Evangelicals

If you are an evangelical who has read this far—or maybe you were curious and jumped right to this section—thank you and congratulations! You understand that climate change is happening, human-caused, and harmful. As an evangelical, you are uniquely positioned to discuss it with friends, family and fellow church members who are unsure about the science—or maybe even think that climate change is a hoax.

Sound like a tall order? Katharine Hayhoe, an evangelical Christian and a climate scientist, has won a national following by explaining how. Author of *Saving Us: A Climate Scientist's Case for Hope and Healing in a Divided World*, a national best-seller, Hayhoe sees no contradiction between her evangelical Christian faith and her commitment to climate science. Rather, she sees them as mutually reinforcing.

"As a Christian, I believe that God created this incredible planet that we live on and gave us responsibility over every living thing on it," she says in a popular TED talk. "And I furthermore believe that we are to care for and love the least fortunate among us, those who are already suffering the impacts of poverty hunger, disease and more."

Born in Canada to parents who were both science educators and Christian missionaries, Hayhoe was surprised when she realized that her husband, a prominent American evangelical pastor, felt he needed to choose between science and faith. It took her two years to convince him otherwise. They then jointly wrote a book: *A Climate for Change: Global Warming Facts for Faith-Based Decisions*, which shows how climate science aligns with conservative Christian beliefs.

Hayhoe points out that evangelical Christians haven't always been hostile to science. The Sixth IPCC Assessment (2021) is dedicated to John Houghton, an evangelical Christian who was chief editor of the first three IPCC reports. Houghton viewed the climate challenge through the lens of his faith. "Looking after the Earth is a God-given responsibility," he wrote before he died in 2020 at age 88 from complications related to COVID-19. "Not to look after the Earth is a sin."

Why are so many Evangelicals skeptical about human-caused climate change? Hayhoe doesn't mince words. "It begins with the fossil fuel companies," she says. "The richest companies in the world have invested in people to tell us... that 200 years of science aren't true." To retain their wealth and power, she explains, the fossil fuel companies tricked ordinary people into believing that addressing climate change would threaten their jobs and lifestyles. Result: "We aren't willing to address the problem but we don't want to be bad people, so we deny that the problem exists," she says.

Solution? "There is one thing all of us can do that we are not doing: talk about it," Hayhoe says. After thousands of such conversations, she has concluded that nearly everybody has the values needed to care about climate. "We just need to connect the dots between the values they already have," she says.

As with other faiths, there's a network for that. Founded in 1993, the Evangelical Environmental Network (ENN) works to "inspire, equip, educate, and mobilize evangelical Christians to love God and others by rediscovering and reclaiming the Biblical mandate to care for creation and working toward a stable climate and a healthy, pollution-free world."

Rev. Mitchell C. Hescox, the president and CEO of ENN, earned his first degree in geochemistry. After graduation, he worked for a company that made coal processing machinery, a job that he says helped prepare him for discussing climate change with evangelicals. He later earned an MA in theology and served as a church pastor for 18 years before becoming the second person to lead the ENN in 2009.

"I felt God calling me to a national ministry," he told me. "After searching my heart and praying, I realized that God put me in this place because I know the energy business." He has also thought a lot about evangelical belief and climate change, co-authoring *Caring for Creation: The Evangelical's Guide to Climate Change and a Healthy Environment*.

Like Hayhoe, Hescox says that the first step in reaching anybody is to open their heart. "You have to be an evangelical to reach an evangelical," he says. "People want to know that you genuinely share their values." In the case of evangelical Christians, this often includes a deeply held opposition to abortion, based on a belief in the sacredness of human life.

Often the most effective opening is to focus on how fossil fuel pollution hurts people's health, especially the health of children, born and unborn, he says. When speaking with older groups of evangelicals, he often cites the links between fossil fuel pollution and conditions such as asthma, allergies, ADHD and autism. "I ask all those who have a child or grandchild who suffers from one of these to stand," he says. "About 70 percent of them stand up."

ENN offers training for evangelicals who want to share their concern about climate change with other evangelicals. "The vast majority of people that we have helped to train are baby boomers, just like myself," he says. "The biggest reason that we have been able to get them involved is helping them to see the impacts of pollution on their children, unborn children and children of any age." The ENN calculates that four million "pro-life Christians" have taken action through the network.

While an evangelical commitment to climate action challenges stereotypes, it's not out of line with the official position of the movement. In 2015 the National Association of Evangelicals, which represents more than 45,000 churches from nearly 40 denominations, urged evangelical Christians to "persuade governments to put moral imperatives above political expediency" on issues of environmental destruction and climate change.

As with other faiths, the door is open to evangelical climate boomers to who want to engage through the lens of their faith. In the case of evangelicals, these voices are especially valuable. Although the number of white evangelicals has declined in recent years, to an estimated 14% of Americans, they are the largest religious grouping in the Republican Party. Just imagine the improved prospects for U.S. climate policies if a significant number of this group became climate activists!

Ecumenical Protestants

Long a formidable force in American culture and politics, ecumenical Protestants—also known as mainline Protestants—greatly diminished in importance during the boomers' ascendency. Mainline protestants comprised more than half of Americans in the early 1960s but many Protestant-born boomers abandoned the church-going habits of their parents, while others joined more lively and conservative evangelical churches.

By 2020, these mostly white Protestant denominations—the best-known being the Episcopalians, Lutherans, Methodists, Presbyterians, and members of the United Church of Christ—constituted only about 16% of Americans, according to the Public Religion Research Institute. Their power and influence declined so much, it no longer made sense to call them "mainline."

In this discussion, I have adopted the terminology of historian David Hollinger, who argues in *Christianity's American Fate* (2022) that the term "mainline Protestants" was already anachronistic in the 1970s, as these traditionally dominant denominations rapidly lost members and cultural standing. He suggests that "ecumenical Protestants" is a better term because of adherents' "willingness to cooperate in ecclesiastical, civic, and global affairs" with Christian and non-Christian groups alike.

Despite their decline in numbers, the historic importance of these denominations still gives their advocacy on climate and other social issues symbolic heft. Most were closely associated with the nation's founding. Their members were the so-called WASPS—White, Anglo-Saxon Protestants—who dominated law, politics, and the upper reaches of America's biggest companies throughout the nineteenth and the first half of the twentieth century.

I was raised in one such denomination, singing in the boys' choir at the United Church of Christ (UCC) in Claremont, California, where my grandparents, YMCA missionaries who spent 30 years in China, were prominent members. In researching this book, I was pleased to discover that the UCC, the religion of my childhood, became the first American denomination to divest from fossil fuels, in 2013.[53]

Like my grandparents, the ecumenical Protestants I have known are broad-minded internationalists, well-read, well-traveled, and committed to making the world a better place. This outlook informs the ecumenical Protestant response to the climate emergency. Although each denomination takes its own positions on social issues, they are closely aligned, coordinating through the National Council of Churches, a politically progressive organization housed in a mid-century office building on Manhattan's Upper West Side known as the "God Box."

In 2013, with membership and budgets declining, the National Council of Churches spun off its Eco-Justice Program into a free-standing entity, the Creation Justice Ministries, which retains ties to the old ecumenical denominations and has expanded to include other Protestant groups, including historically Black denominations. The Creation Justice Ministries Board of Directors reads like a Who's Who of ecumenical Protestant leadership. Each month the Creation Justice Ministries issues a call to action, usually a letter-writing campaign. In August 2022, for example, the call was to urge the House to pass the climate-focused Inflation Reduction Act.

While the Creation Justice Ministries provides a valuable coordinating mechanism for ecumenical Protestant leadership, for lay boomers with roots in these denominations, the best path towards faith-based climate advocacy is to find a church in your community that is already engaged in this work, or to search online for statements and teachings on climate within your denomination. In 2021, for example, the Episcopal Church sent 24 delegates to the UN Climate Conference in Glasgow, "to learn about the state of the climate crisis and efforts to address it, and to bring what they learned back to the wider church."

Many ecumenical Protestant churches are active in interfaith climate action efforts, so connecting with a local church's climate advocacy work can open the door to engage with faith-based climate activists belonging to other traditions.

Black Churches

For more than 300 years, Black churches in America have provided a haven for Black people forced to struggle daily against systemic racism. Besides their role as places of worship, Black churches have long served as places of racial solidarity and civic activity.

Black churches played a central role in the Civil Rights movement. One of the leading civil rights organizations, the

Southern Christian Leadership Conference, was founded in 1957 at the Ebenezer Baptist Church in Atlanta, Georgia, by nationally known Black pastors including Dr. Martin Luther King. The SCLC's leaders believed that churches should be politically involved and held many of their meetings at Black churches, which became important symbols in the battle for civil rights.

Today the term "Black church" is widely understood to include seven major Black Protestant denominations: the National Baptist Convention, the National Baptist Convention of America, the Progressive National Convention, the African Methodist Episcopal Church, the African Methodist Episcopal Zion Church, the Christian Methodist Episcopal Church and the Church of God in Christ. Taken together, African American denominations account for about 7% of Americans who identify with a faith group.

It is now well known that people of color, especially African Americans, are exposed to industrial pollution from fossil fuels more often than others and are more likely to suffer from cancer, lung and heart diseases, and premature death as a result. Polluting industries tend to locate in poor communities that lack the power to block them, especially majority Black communities that have been weakened by redlining, lack of investment, and other aspects of systemic racism. This terrible disparity was heightened during the COVID-19 pandemic, when pre-existing health conditions contributed to a significantly higher mortality rate among Black Americans.

Black churches, like other institutions in the African American community, have long known about these injustices and struggled to overcome them. In recent years, the wider U.S. environmental and climate action movement has belatedly taken note, paying greater attention and in some cases providing funding and other support to Black-led organizations fighting for environmental justice. Black church leaders point out that their institutions can often be an effective and valuable conduit for these funds.

Green the Church has been in the forefront of efforts to link the Black Church and the environmental and sustainability movement. Founded in 2010 by Rev. Ambrose Carroll Sr., a pastor in Oakland, California, Green the Church aims to expand the role of Black churches as "centers for environmental and economic resilience."

Green the Church works through three pillars: "amplifying green theology," working with church leaders to develop and share environmental teachings; "supporting sustainable practices," such as helping churches to save money through energy audits and rooftop solar; and "building power for political and economic change," by supporting member churches engagement in government policy decisions "to transform how our government acts on climate change, supports the green economy, and invests in resilient communities."

Rev. Carroll invites Black church leaders, boomers or otherwise, who want to explore these opportunities to reach out to Green the Church. He also welcomes volunteers who want to work within the organization itself in such areas as administration, outreach and marketing. The website has a form for those who want to apply.

"As climate change accelerates, it becomes clearer each day that those who will suffer first and most will be Black Americans," Rev. Carroll wrote in an October 2020 appeal to Black churchgoers to turn out for the November election. There were many reasons for Black people to go to the polls, he said, including showing support for racial justice, for a science-based response to the COVID pandemic, for an equitable society and a living wage, and for ensuring that access to quality healthcare is a right.

Protecting God's creation, the Earth, also belongs on that list, he wrote, citing a poll showing that the vast majority of Black

Protestants favor climate action. "What this recent poll also shows me is that many Americans, particularly Black Americans of faith, will also vote to take responsibility as guardians to the greatest gift that God has bestowed upon us," he wrote.

A millennial, Rev. Carroll fondly remembers the young boomers who were his junior high school teachers as idealistic people who were part of a multi-racial coalition. "I've got pictures of white folks listening to the Panthers," he told me. Later, he said, some of these boomers "cut their hair and put on suits," abandoning social justice struggles. He is hopeful that the climate justice struggle will lead boomers of all races "to become reacquainted" with the values of their youth.

Support for Black churches in retrofitting their buildings could be one area for cooperation, he says. Church buildings are often the Black community's largest shared asset, but they are threatened by gentrification and need to be upgraded. "All those building need to be retrofitted, to make sure that they have clean air and clean water," he told me. Such retrofitting can also provide job training for Black young people in new clean, green jobs, he added.

Muslims

Unlike other U.S. religious groups, most U.S. Muslim adults (58%) are immigrants, the result of a 1965 law that lowered barriers to immigration from Asia, Africa and other areas outside Europe. The U.S.-born American Muslim population is diverse and includes descendants of immigrants, converts to Islam, and the descendants of converts.

Estimates of the number of Muslims in the United States vary from about 3 million to 5 million, or about 1.5% of the U.S. population. Muslims tend to be younger than other religious groups, with a median age of 33 years old, compared to 47 for other faiths. Boomers account for just 15% of American Muslims, about half the percentage for other religious groups.[54]

Muslim advocacy for climate action reflects these unusual demographics: a mostly young population that shares the concerns of other American youth about climate change while also dealing with the more immediate concerns facing recent immigrants, including anti-Muslim sentiment among some segments of the U.S. population.

Internationally, Muslim views on climate are shaped by the highly unequal distribution of the benefits and costs of fossil fuels among majority Muslim nations. Some Muslim countries have grown immensely rich from oil. Others, such as Pakistan, the world's second most populous Muslim nation, are suffering horribly from climate impacts.

From June to October 2022, melting glaciers and heavier-than-usual monsoon rains inundated a third of the country. In January 2023, UNICEF said that four million Pakistani children living near contaminated flood waters were "fighting a losing battle against severe acute malnutrition, diarrhea, malaria, dengue fever, typhoid, acute respiratory infections, and painful skin conditions."

The highest-level Muslim call for climate action is the *Islamic Declaration on Global Climate Change* created at a 2015 international symposium in Istanbul attended by academics, religious authorities, and civil society leaders from 20 countries ahead of the Paris Climate Conference. Drawn from a body of ethical writings based on the Holy Quran, the declaration asserts that human beings are God's *khalifah* or "steward" on Earth and must therefore care for God's creation.

The declaration is surprisingly forthright on the question of fossil fuels. "We particularly call on the well-off nations and oil-producing states to lead the way in phasing out their greenhouse gas emissions as early as possible and no later than the middle of the century," it says. In addition, clear emissions reductions targets and monitoring systems should be put in place, and rich countries should provide "generous financial and technical support" for poor countries to help wean them off fossil fuels.

The declaration informs the work of the Islamic Society of North America (ISNA), the largest such association in the United States. ISNA's "Green Initiative" works to raise awareness about the "catastrophic effects of climate change," to reduce waste like plastic water bottles and Styrofoam, and to promote the use of solar energy to reduce the use of fossil fuels. The overall goal is "to promote environmental and social justice."

The initiative is led by Imam Saffet Abid Catovic, ISNA's head of interfaith, community and government relations and a member of the Parliament of the World's Religions Climate Action Task Force. At the UN Climate Conference in the Egyptian resort of Sharm El-Sheikh in November 2022, he spoke loudly and clearly against fossil fuels. "We need no more deaths as a result of the fossil fuel proliferation," he told a small group of demonstrators in an officially sanctioned protest zone. "Our leaders, our negotiators, need to know: no more business as usual."

I watched that demonstration on YouTube. The month before, I watched in person as Imam Catovic was arrested at the White House together with a rabbi and an ecumenical pastor as part of a mass civil disobedience action calling on President Biden to declare a climate emergency and end new fossil fuel projects.

Another Muslim group striving to address the climate emergency is Green Muslims, a volunteer-driven non-profit based in the suburbs of Washington, DC, which engages locally and works to connect Muslims across the country to nature and to environmental activism. The group hosts educational and outdoor events and serves as a bridge between the Muslim community and local climate action organizations.

Executive Director Sevim Kalyoncu grew up in Alabama surrounded by woods and creeks where she says she experienced her most direct connection to God. After earning degrees at

Georgetown University and the University of Chicago, she settled in the Washington, DC, area where she became a mother and began sharing her love for nature and her concern about climate change with other Muslims.

"I wanted my children and other children to have the same opportunity I did to connect with God through nature," she told me. She began organizing nature walks and other outdoor activities and sometimes participated in interfaith climate activities. She noticed that many of the faith-based groups benefitted from having older supporters.

"I was envious of their boomers," she told me. "The boomers have time and energy and resources, even money, to support the work." As the head of Green Muslims, she has worked hard to attract older members, efforts that she says are beginning to yield results. "We are attracting more and more people from older generations," she says.

Kalyoncu encourages Muslims of all generations to use the online resources of the ISNA Green Initiative and Green Muslims. Her advice to Muslim boomers concerned about climate change: get in touch with your local community, find out who else shares your concerns, and join or start a Muslim "green team." Join in outdoor education activities, such as learning about and planting native plants. Sign up for the Green Muslims newsletter and consider supporting the organization, by donating, volunteering, or both.

Jews

I am a Jew by choice, having studied at a Northern Virginia synagogue with Rabbi Laszlo Berkowits, a Hungarian-born Holocaust survivor who lost most of his family in the Nazi death camps. His enthusiasm for life after having suffered such a searing experience was one of the things that drew me to Judaism. Another factor was Judaism's deep commitment to social justice, one grounded in the many biblical commandments to welcome the stranger and to care for the poor. This commitment underpins a growing Jewish movement for climate action.

Like other religious groups, Jews have turned to ancient texts seeking guidance to address an emergency that ancient people could hardly have imagined. American Jews speaking about climate change often cite a story from Ecclesiastes about what God told Adam in the Garden of Eden: "Look at My works! How beautiful and praiseworthy they are. Everything that I have created, I created for you. Take care not to damage and destroy My world, for if you destroy it, there is no one to repair it after you."

Jews are a small proportion of the U.S. population—about 2.4% of the total population or 5.8 million people, counting 1.5 million who identify as "Jewish" but say they are atheist, agnostic, or have no religion in particular.[55] Even so, we have been active and visible in many social struggles.

In the 1960s, American Jews actively supported the Civil Rights movement. Rabbi Abraham Joshua Heschel, a leading theologian, and Rabbi Maurice Eisendrath, the leader of Reform Judaism, marched beside Dr. Martin Luther King, Eisendrath famously carrying a Torah. The example of bringing sacred objects into the struggle for justice inspires Jewish climate activism today.

Several Jewish groups campaign for environmental justice. The Coalition for the Environment and Jewish Life (COEJL), established in 1993, works to build awareness and mobilize advocates. Hazon, established in 2000, sponsors bike rides and awards a "Hazon Seal of Sustainability" to synagogues. Newer groups include the Jewish Earth Alliance, which organizes volunteers to engage with Congress, and the Jewish Climate Action Network, a loose affiliation of city and state-based groups.

Jewish climate advocacy gained fresh momentum in 2020 with the launch of Dayenu, a Jewish call to climate action.

"Dayenu" means "enough" and features in a song we sing at Passover thanking God for leading the Israelites out of Egypt. In this case, "Dayenu" has two meanings, conveying that there is enough in the world to meet human needs as well as anger at those who refuse cries for environmental justice—as in "enough already!" Dayenu supports advocacy and "spiritual adaptation," helping Jews come to terms with climate grief and loss, while cultivating the hope needed to struggle for social and environmental justice.

"We know that many people are living with fear, anxiety, guilt and the sense of disconnection that comes from going about one's day-to-day business with the looming—or already present—threats to the safety and well-being of themselves and those they love," says Dayenu founder, Rabbi Jennie Rosenn. "Dayenu seeks to support Jews and Jewish communities to spiritually adapt to this new and terrifying reality."

Dayenu's leadership includes Phil Aroneanu, a co-founder with Bill McKibben of 350.org, and Vicki Kaplan, former organizing director at MoveOn.org, a progressive advocacy group. The group's national organizer is a former volunteer with Sunrise Bay Area in California, one of the group's most active hubs. That former Sunrise volunteer is my daughter, Muriel MacDonald.

I would have been involved in Jewish climate advocacy whether or not Dayenu existed—and whether or not my daughter worked there—because I believe that the teachings of my adopted religion demand it, and because I believe that faith-based advocacy is one of the most effective things I can do to address the emergency.

But Dayenu has made it easier, more fun, and more effective. As the co-chair of the Dayenu Circle at my synagogue, I have helped organize two actions, one in Virginia encouraging Sen. Mark Warner to do more to win passage of President Biden's Build Back Better legislative package, and one in Washington,

DC, urging the "*schmutzy* (dirty) seven" big banks and financial institutions to stop investing in fossil fuels.

For Jewish boomers who want to become climate activists, Dayenu offers tips for organizing a new Dayenu Circle, training, and periodic campaigns underpinned by strategic analysis and made lively with creative use of Jewish symbols. Examples include the *shofars* (rams' horns) we blew outside Sen. Warner's office to call attention to the urgency of the crisis, and the *matzoh* (unleavened bread) we held aloft in Washington, DC, fossil fuel divestiture action as we chanted: "Move Your Dough!"

Interfaith Action

We began this chapter explaining how interfaith action can be a powerful tool to draw attention to the climate emergency. For those engaged in faith-based climate advocacy, joining in interfaith actions is a great way to leverage your knowledge and networks. You needn't be an ordained religious leader or hold some other official position to do this.

In the summer of 2021, I traveled to northern Minnesota to participate in an interfaith delegation supporting Native Americans and others opposing the Line 3 pipeline. Native American leaders had invited climate activists and people of faith to join them in a mass protest along the pipeline route.

The interfaith group stayed three days at a church camp on the banks of a beautiful lake, where we sang, prayed, and heard from Winona LaDuke and other Native elders about their struggle to protect their land and water, including the headwaters of the Mississippi. We were an intergenerational group of about 100 people, young climate activists with body piercings, tattoos and preferred pronouns, and boomers in Birkenstocks, some joining a climate protest for the first time.

The morning of the march, members of half-a-dozen faiths shared prayers and explained why their faith led them to oppose Line 3 and other fossil fuel infrastructure. Only a few spoke in an official capacity; most were simply active members of their faith. During the march we carried signs identifying us as members of various faith communities. Press reports later noted the participation of religious groups, making clear the broad-based nature of the opposition to the project.

Fletcher Harper, an Episcopal priest and the executive director of GreenFaith, which helped to organize the Line 3 interfaith delegation, believes that boomers are key to tipping the balance towards ambitious action.

"The boomer generation cut its teeth at a transformative time in American history in the '60s and '70s," he told me. "There's a lot at stake. Boomers have every reason to want to leave a legacy that they can feel good about." We are either already retired or far enough along in our working lives that we can afford to take some chances, he explains. "We need that kind of unfiltered energy and we need people who have been around the block."

Like Novey of Interfaith Power and Light DMV, Rev. Harper and other faith-based climate organizers I spoke with all recommend finding like-minded people, whether by joining an existing group or starting one, rather than trying to persuade those who don't share your values or concerns.

"Minimize the amount of time you spend barking up useless trees," Rev. Harper advises. "Do a kick-off event at your congregation. Form a local circle. People will come up to you and self-identify. You want to find others who care and are concerned and are fed-up with the lack of action."

For those getting started, GreenFaith offers referrals to existing groups, and training, coaching and support for those who want to start their own climate action circle. You can start by signing up for the GreenFaith newsletter and filling in a simple form on their website. Rev. Harper says it's fine for religious institutions to start with simple steps, like improving energy efficiency and promoting plant-based diets, while encouraging their members to take similar steps at home. But he adds that religious communities mustn't stop there. People of faith "must use our moral power to press for policies that move us in the direction that we need to go," he says.

What's Next?

For boomers who want to make a difference, doing what we can in in our personal capacity, and then tapping the power of our knowledge and networks, joining a volunteer organization like Th!rdAct or Citizens' Climate Lobby, and getting involved in faith-based advocacy can help a lot.

These actions alone, however, won't be enough to achieve the rapid social and economic transformation that is necessary to avert a planetary catastrophe. What more can be done? And how can members of our generation help make it happen?

The next chapter offers an introduction to civil disobedience and non-violent direct action. As we shall see, these tactics are deeply rooted in American tradition—and in the formative experiences of our generation. Sharing my own experience, I will invite you to consider becoming part of the growing community of boomers who participate in or otherwise support non-violent civil disobedience as a means to address the climate emergency.

Chapter 6: What's Faith Got to Do with It?
Action Checklist

- Learn about faith-based teachings on climate
- Tap national organizations for advice and support
- Join in faith-based climate advocacy
- Encourage your clergy and other faith leaders to speak out
- Explore inter-faith action

Should I get arrested?

Civil disobedience is baked into the American psyche and millions of boomers say they are willing to risk arrest to protect the climate. Should you be one of them?

Chapter 7

Should I get arrested?

We have seen that American boomers called the shots during a crucial period of planetary history, that we failed to put our country on the path to a low-carbon economy, and that this matters profoundly, not only for the United States and our children and grandchildren, but for the entire world and countless generations to come. We also have seen that our generation still has enough power to help avert catastrophe. We identified six impactful actions that we can take on our own, and we explored how we can use our knowledge and networks, working with others, including people with a shared faith tradition, to have an even bigger impact.

All well and good. But we know from the continued rise in emissions, and the resulting surge in record high temperatures and climate-related disasters, that the window of time when incremental change could have averted catastrophe has slammed shut. Without a very rapid transformation of energy systems and societies, the world will fail to avert a catastrophe that will last for millions of years. So, you might be asking yourself: what else can I do? The answer, for a growing number of Americans—including millions of boomers—is participation in civil disobedience and non-violent direct action.

Before we go further, a quick glossary. In public discourse, the terms "civil disobedience" and "non-violent direct action" are often used interchangeably. Even among political scientists, definitions of these and related terms differ by author. Activist John Halstead reflects one common view, defining "non-violent civil disobedience" as intentionally risking arrest to make a *symbolic statement.* Getting arrested in front of the White House falls into this category. Halstead defines "non-violent direct action" as trying to achieve a goal directly, such as shutting down a coal plant with a blockade. Both non-violent civil disobedience and non-violent direct action are forms of civil resistance, a broader term that also includes protests and marches.

In the discussion that follows, I've generally applied Halstead's definitions, while sometimes adopting the more common approach of using the two terms interchangeably. For example, when analyzing the findings of *Climate Change in the American Mind* surveys, I assume that respondents were unfamiliar with these distinctions and that their professed willingness to participate in what the survey called "non-violent civil disobedience" applies to both types of activities.

Regardless of how one defines the terms, support for civil disobedience to push for climate action is surprisingly high among Americans. A 2021 *Climate Change in the American Mind* survey found that about 5% of Americans "definitely would" participate in non-violent civil disobedience against corporate or government activities that make global warming worse "if someone they like and respect asked them to do so."[56]

Among those who are "alarmed" about climate change, the percentage was much higher, with 19% saying they "definitely would" participate and 29% saying that they "probably would" participate. As climate impacts have become more frequent and extreme, the proportion of Americans who say that they are "alarmed" about climate change has surged, to almost one-out-of-three in 2021.[57] Assuming these trends continue and that the association between being "alarmed" and being willing to risk arrest holds, the number of Americans who are prepared to participate in civil disobedience to press for climate action seems likely to rise further.

According to the *Climate Change in the American Mind* survey, 8% of Americans who are boomers and older said that they "definitely would" participate and 14% said that they "probably would" participate in non-violent civil disobedience to oppose actions that make climate change worse. While this was lower than for millennials (14% "definitely" and 21% "probably"), it was about the same as for Gen X. Given that there are about 70 million boomers alive today, this implies that almost six million of us say we "definitely would" participate in such actions if asked by somebody we like and respect. Even discounting for a certain amount of bravado in such surveys, there are likely still millions of us who are prepared to take this step, a huge untapped potential for climate-related boomer civil disobedience.

If millions of boomers and others who say they are prepared to engage in climate-related civil disobedience actually did so, would it make a difference? Yes! In a now-famous 2012 study, *Why Civil Resistance Works: The Strategic Logic of Nonviolent Conflict,* political scientists Erica Chenoweth and Maria J. Stephan analyzed hundreds of movements for political change that occurred between 1900 and 2006.[58]

They concluded that non-violent protests are twice as likely to succeed as armed conflicts—and that movements that actively engage at least 3.5% of the population "never failed to bring about change." The data from the *Climate Change in the American Mind* surveys suggest the United States has plenty of scope to pass this threshold for climate—if enough of us, including us boomers, match our words with action.

Our generation should be in the vanguard. Having come of age when civil disobedience was transforming America, we know a thing or two about disruptive tactics. Some of us who are older participated ourselves, while middle and younger boomers saw such actions on the nightly news. From the 1960s onwards, civil disobedience helped end the Vietnam War, achieve landmark environmental legislation, and end the Jim Crow laws that provided the legal underpinnings for institutionalized racism.

While we boomers didn't do any of this alone, our participation was key to each of these changes, and often we led the way. Can we recapture this knack for disruption in service of passing on a livable planet to future generations? In doing so, we would not only reconnect with the idealism of our youth; we would also become participants in an American tradition that began well before the 1960s.

Civil Disobedience: An American Tradition

Civil disobedience—risking arrest to shine light on injustice—is baked into the American psyche, starting with Henry David Thoreau's famous 1847 essay *On the Duty of Civil Disobedience.* Thoreau argued that citizens have an obligation to uphold justice, even if this means violating unjust laws and bearing the consequences—including jail or prison. His disobedience—refusing to pay a poll tax—was grounded in his opposition to slavery and the U.S. invasion of Mexico. In language that feels startlingly up to date, he compares government to a machine producing injustice. Citizens have an obligation, he writes, "to stop the machine."

The essay deeply influenced some of the world's most famous struggles for social justice. Almost 100 years later, Mahatma Gandhi wrote that Thoreau's example was "exactly applicable" to the non-violent struggle he was then leading for India's independence from Britain. Dr. Martin Luther King wrote that "Thoreau came alive in our civil rights movement… Whether expressed in a sit-in at lunch counters, a freedom ride into Mississippi, a peaceful protest in Albany, Georgia, a bus boycott in Montgomery, Alabama, these are outgrowths of Thoreau's insistence that evil must be resisted and that no moral man can patiently adjust to injustice."

Civil disobedience was also central to opposition to the Vietnam War, when young boomers publicly burned their draft cards, courting arrest and prison. As a teenager, I remember hearing that Russell, the husband of my cousin, Cynthia, had burned his draft card and was going to jail as a result. Cynthia's parents, my very conservative aunt and uncle, were furious, arguing that he exposed himself to unnecessary risk. Cynthia enrolled in law school to help with Russell's legal appeals and later became a public defender. The marriage did not survive Russell's five years in prison and he ceased to be part of our extended family. But, to me as a teenager, Cynthia and Russell were heroes.

Such 1960s civil disobedience has inspired the climate movement, with Bill McKibben comparing the mass arrests outside the White House in 2014 protesting Keystone XL to mass protests against the Vietnam War. Since then, climate activists have engaged in hundreds of acts of civil disobedience across the country, and hundreds of people—possibly thousands—have been arrested. These acts helped to raise awareness about the climate emergency and sometimes delayed or even stopped major fossil fuel infrastructure projects.

What does civil resistance to block climate destruction look like? How does it feel? What are the risks, and how can these be managed? While I am far from an expert, I am proud to have been arrested three times in non-violent climate protests—likely more than many of my readers, certainly much less than many of the experienced activists I've come to know and admire. When people ask me how many times I've been arrested, I like to reply with a story I heard long ago about an elderly Civil Rights activist. Asked how many times she had been arrested, she replied: "not enough!"

Following are several examples of climate civil resistance involving boomers, five that I learned about from others and two where I was personally involved. The chapter ends with some reflections on the potential and the challenges of using these tactics to sound the alarm and demand action to address the climate emergency.

Boomer Activists and the Mountain Valley Pipeline

As the climate threat becomes more dire, the frequency and creativity of climate-related civil disobedience actions are growing. Millennials are often in the forefront of the most dramatic actions, such as the "tree sitters" who stay aloft in tree-top shelters, sometimes for months, to block new fossil fuel infrastructure. Although the physical demands of tree sits and other blockades can make them a poor fit for most boomers, older people often help in supporting roles, such as preparing meals, looking after children, serving as legal observers and media liaisons, and providing financial support. More often than you might think, however, boomers also put themselves at risk.

One example is the record-setting Yellow Finch tree-sit that blocked construction of the Mountain Valley Pipeline (MVP), a 303-mile fracked methane gas pipeline from West Virginia to southwestern Virginia, for an astonishing 932 days. Activists organized by Appalachians Against Pipelines took turns camping on platforms 50 feet in the air, supported by teams on the ground that provided food and supplies. It ended in March, 2021, when a court authorized police to remove the ground encampment and the last two tree sitters were extracted with a hydraulic crane. Each was sentenced to months in jail and fined thousands of dollars, as well as being ordered to pay the MVP nearly $150,000 for the cost of removing them.

Most of the tree sitters were young people who shielded their identity from the media to avoid hurting their future employment prospects. Scott Ziemer, a 69-year-old grandfather, sailing instructor and one-time arborist had no such concerns.

"I've worked most of my life outdoors. I know about ropes and working at heights, so I was training young people for the Yellow Finch blockade," Ziemer later told me. "One day I realized that if they were willing to take this risk, I should be, too."

In April 2019 he spent a week atop a tree platform at Yellow Finch and briefly became the public face of the blockade. "By occupying a tree sit in the path of the Mountain Valley Pipeline, I am adding my voice to those who are fighting to slow down and stop the burning of fossil fuels, which are the primary cause of climate change," he said in a Facebook post by Appalachians Against Pipelines. "We're so excited that Scott has joined us at the Yellow Finch sits, and that he is harnessing his privilege by putting his body on the line to continue blocking MVP construction," the unsigned post said.

Three months after the Yellow Finch tree sit ended, three boomers locked themselves to and inside a broken-down Crown Victoria on a back road to different portion of the MVP route, blocking access to the construction site. The vehicle was partly filled with cement, to make it harder for police to remove, and brightly painted with climate-action and anti-pipeline slogans, including: "Old Hills & Old Folks Resist." About 20 people rallied in support, some hiding in the nearby woods to take pictures.

Deborah Kushner, a 66-year-old retired mental health worker who would go on to become a co-founder of Th!rdAct Virginia, sat on the trunk in a rocking chair, shielded from the sun by an umbrella, her left arm chained to the car. In the pictures, she looks calm and determined. "My presence here today is just a small gesture of dissent. It's what I can do to delay—even for a few hours—the march towards assured oblivion," she said before her arrest. "My action today is very much out of my comfort zone. I'm compelled to act because we are out of time. My stand today is for a future in which we all can thrive."

Alan Moore, a 57-year-old environmental scientist with diabetes requiring dialysis, asked to be welded inside the vehicle before police arrived, so it would take them longer to dismantle the blockade. "Many who have acted in opposition to the insanity of a twenty-first-century fracked gas pipeline have been attacked, arrested, brutalized, and dragged into a costly and dehumanizing legal system," he said. "Each of us deeply feels the ache of living in the Appalachian sacrifice zone."

Moore told me later that he felt compelled to support young people opposing the pipeline who had been treated badly by police. "You could see the change in their behavior when we showed up," he said, adding that the police seemed to treat older people better.

The third member of the trio was Bridget Kelley, a 63-year-old central Virginia resident. "I have done everything in my legal power to stop this pipeline from the destruction it wields, but all to no avail," she said. "We have to stand up for our rights. Come and join us! Together we can and will win."

Since those actions the MVP has gone from being a local issue to a national symbol of the struggle to block new fossil fuel infrastructure. West Virginia Sen. Joe Manchin extracted from Democratic Party leaders a promise to address so-called "permitting reform" and to remove obstacles to the completion of the MVP in exchange for his support of a stripped-down version of the Build Back Better package re-branded as the Inflation Reduction Act (IRA). Protesters dubbed the backroom agreement "The Dirty Deal."

With the MVP years behind schedule and millions of dollars over budget, completion is uncertain. As with Keystone XL, Dakota Access Pipeline, and Line 3 in Minnesota, the MVP will be a bellwether in the ongoing struggle to block new fossil fuel infrastructure, thereby hastening the end of the fossil fuel era. Whether these efforts succeed will depend partly on how many of us boomers are willing to risk arrest to disrupt business as usual.

Rocking Chair Rebels, Raging Grannies, 1,000 Grandmothers

Several groups of boomers have organized actions that use their status as "senior citizens" to grab attention. The Walk for Our Grandchildren is one such effort. At the first walk in 2013, a multi-generational group led by older Americans walked from the presidential retreat at Camp David, Maryland, to the White House to tell President Obama to end fossil fuel extraction.

In June 2021, the group walked from President Biden's hometown of Scranton, Pennsylvania, to Wilmington, Delaware. This time they courted arrest by sitting in rocking chairs to block a street in front of the JPMorgan Chase credit card headquarters demanding that the bank stop funding fossil fuel extraction. Dubbed the "Rocking Chair Rebels," they were arrested and fined $10 each for blocking traffic. Instead of paying, they went to trial, arguing that raising the alarm about the climate emergency justified blocking the road.

At trial they presented as evidence the 50-page summary of the most recent IPCC report and the 2021 *Banking on Climate Chaos* report showing that the world's 60 largest banks—including JPMorgan Chase—poured $3.8 trillion into fossil fuels from 2016–2020. The Rocking Chair Rebels lost the case and were charged about $100 each in court costs. They viewed entering the documents into the court record as one small step in the larger campaign to sound the alarm.

"People all over the country are doing the same things in their states," said Ted Glick, one of the Rocking Chair Rebels. As a young man, Glick spent nearly a year in prison for his opposition to the Vietnam War. As an old boomer he has been arrested more than two dozen times in non-violent civil disobedience to advance climate justice. "All of us have to contend with a federal government and mega-banks that have continued to allow and finance new fossil fuel projects, digging the climate crisis hole we are in even deeper," he said.

Another elder-led group that has engaged in civil resistance is the Raging Grannies, a Canadian-founded group of women who write new lyrics for well-known tunes, then dress up in clothes that poke fun at old lady stereotypes and sing at protests. The Raging Grannies have been the subject of two films and two books.[59]

A group called 1,000 Grandmothers was founded in 2016 in Oakland and Berkeley, California, to support the Grandmothers at Standing Rock, a group of Indigenous women water protectors attempting to block the Dakota Access Pipeline. 1,000 Grandmothers has since engaged in non-violent direct action as well as organizing a successful fundraising concert.

The group's principles include: "We create and support opportunities for elder women to step up courageously to acts of civil disobedience if they are motivated to do so. At the same time, we are committed not to create a macho culture that promotes the notion that doing civil disobedience is more important than doing every other thing that grandmothers choose to do to address the climate crisis."[60]

Boomer non-violent direct action to disrupt new fossil fuel infrastructure need not involve lots of people. In July 2021, three Ohio grandmothers blockaded a Line 3 pipe storage yard, locking themselves to the gate and to each other. Their action in support of Indigenous activists disrupted operations for three hours and helped draw additional attention to the risks associated with the pipeline.

"This is my time. It's my generation that made the mess... We want to take care of our children and our grandchildren but we did such a poor job of not cleaning up the mess," said Judy Smucker, one of the three. "I can do it now. There's nothing to stop me. We're a force––grandmothers are a force, a real force. We can be if we unite."

Getting Arrested at the White House: Keystone XL

While my own experience with civil disobedience is paltry compared to that of many veteran activists, it may be of interest to readers who are new to this idea, perhaps considering for the first time joining in or supporting such activities. My first two arrests occurred outside the White House in well-organized actions in opposition to the Keystone XL Pipeline.

The first, in 2011, was led by McKibben and 350.org. Those of us who were considering risking arrest were required to participate in a training session held in the basement of a nearby church the evening before. We were each paired with a buddy, told exactly what to expect, and walked through simulations of various scenarios, such as how to deal with hecklers (ignore them).

The trainers told us we would risk arrest by standing in front of the so-called "postcard window," the spot where tourists snap pictures of the White House. Park Police would warn us three times to move, then handcuff us, put us in a police van, and drive us to the jail where we would be fined $100 each and released. If we stayed out of trouble for 90 days, the misdemeanor charges of obstructing a sidewalk and refusing a police order to move would be expunged from our records. We were told to bring our I.D. and $100 cash, pay the fine and walk free.

Events the next day unfolded exactly as we had been told. Although the risks to our well-being were very small and well-managed, being arrested, handcuffed, driven through the city in a police van, and booked into jail, even if only for a brief time, was nonetheless a moving experience. It felt a bit like a mixture of participating in a religious ritual and performing in a play. I was proud to be standing up for something important.

I've since learned that others who experience a civil disobedience arrest for the first time feel much the same. Peter Kalmus, a NASA climate scientist who has been outspoken about the importance of scientists joining in civil resistance, was arrested for the first time in April 2022 when he and other scientists chained themselves to a Chase Bank entrance in Los Angeles. Kalmus said in an interview after the arrest: "It really does feel incredible. It feels so liberating and the right thing to do."

Asked for his advice for others risking arrest he said: "See yourself as a vessel for this message which you know so deeply is correct... to stand up for life on this earth. That's what gives me the courage to do this. The damage is going to reverberate through the earth's geological history for millions of years.... Getting arrested is an expression of love, energy you are putting out into the world without expecting anything in return."[61]

My second arrest was a year later, also at the White House, in an interfaith protest designed to keep up pressure on the Obama administration to stop Keystone XL. Events again unfolded as planned, except that moments after I was released from the jail, the nation's capital was shaken by the biggest earthquake in more than 100 years. Although damage to the city turned out to be minor, for me the quake heightened the drama of a memorable day.

Coal Baron Blockade

My third arrest was different. Like many Americans, I had watched with mounting frustration as Build Back Better passed the House but was blocked in the Senate by an implacable GOP and two conservative Democratic senators, Kyrsten Sinema of Arizona and Joe Manchin of West Virginia. Sen. Manchin emerged as key and negotiations with him dragged on for months, as momentum slipped away.

That Manchin had grown rich off West Virginia coal was widely known but the details were unclear until *The New York Times* published an expose showing how Manchin had used his influence while serving as governor to get permits to burn a highly toxic type of coal waste known as gob, short for "garbage of bituminous." He sold the otherwise worthless gob to the Grant Town power plant.

The article said that figuring out who owned the Grant Town power plant was "a bit like handling a set of Russian nesting dolls" due to complex arrangements among multiple holding companies. It unpacked these obfuscating ties to show that selling gob to the plant yielded some $500,000 a year for Manchin and his family, who made additional money by selling high-priced electricity to his constituents, many of whom live in poverty and suffer from illnesses related to the coal industry.

Furious at these revelations, I was more than ready to respond when I saw a Tweet from a group called West Virginia Rising recruiting people to participate in a "Coal Baron Blockade" of the plant. I hoped the action would draw attention not only to Manchin's role in blocking climate legislation but also to the outrageous conflict of interest inherent in his serving as the chair of the Senate Energy and Natural Resources Committee while earning half-a-billion dollars a year burning one of the dirtiest forms of coal.

I signed on and was told to travel to a church camp near Grant Town, a five-hour drive from my home in Northern Virginia, where I arrived the afternoon before the action. We were briefed on the goals—to call attention to Manchin's conflict of interest by "blockading" the plant—and asked to sort ourselves into three groups: green, for those who would support the action but not risk arrest; yellow, for those who were willing to accept some risk of arrest; and red, for those fully prepared to risk arrest. Having participated in the Line 3 protest in northern Minnesota as part of the interfaith delegation's green group, I felt ready to do more. I decided to join the red team.

After a dinner of standard climate camp fare—rice, vegetables, pasta—organizers reminded the 200 or so people at the camp of our commitment to not engage in violence and to not damage property, refraining from even relatively minor acts such as "tagging" surfaces with graffiti. Twice we pledged to observe these rules by giving a thumbs-up, once for not hurting people, once for not damaging property. Each time we were asked to gaze around the room, mutually witnessing the thumbs-up pledges.

The plan, we learned, was for one group to block the main entrance to the plant while a smaller group would hang a banner that read: "Manchin: Stop Burning Our Future" at the back gate. I decided to join the back gate group. Everybody, regardless of their anticipated role, filled in a "jail support" form providing the names and contact information of people to be informed if we were arrested, standard practice in civil disobedience actions.

After the briefing, people headed off to sleep, some in bunk rooms, others in tents. I pitched my tent on the wet grass behind the dining hall and slept through the night to the sound of a slushy sleet falling on the tent. In the morning we had a simple breakfast and met with our "affinity groups."

Affinity groups, a common organizing tool in large-scale civil resistance, are small groups of people who act as a team. This can include people trained for specific roles, such as "legal observers," who document what happens, "marshals," who show fellow protesters where to go, and "street medics" who look after protesters' well-being. Although the last sounds dramatic, and there surely are exceptions, in my experience "street medics" mostly encourage folks to apply sunscreen and drink plenty of water, as well as sometimes providing basic first aid.

After the meetings, we boarded mini-buses and headed to the plant. On arrival, we climbed off the buses and started walking up a dirt road, waving signs and chanting. I found myself in a line of about six people at the front of the group. About a dozen police officers blocking the road ordered us to stop.

"You are trespassing," one officer shouted. "Stop now!" I planted my feet, not attempting to go forward but also not backing away. "You are under arrest. Turn around," the officer in front of me shouted. I did as I was told and was handcuffed. I saw police arrest two other people both much younger than me, a fellow protester at the front of the line and a photographer documenting the protest.

It was about 1:00 in the afternoon. The three of us spent the next 12 hours in detention, being driven first to a nearby fire station; then to a police station where we were photographed, fingerprinted and booked; and finally to the county jail, where we were again photographed, fingerprinted, and booked. Besides us, the small cell held about five other men, mostly in jail for drug and alcohol offenses. We sat and lay on dirty foam mattresses that lined the small cell wall-to-wall.

Sometime after midnight each of us was taken from the cell and arraigned via speaker phone, with a lawyer and a judge, unseen, on the other end of the line. That's when we learned we were being charged with felonies, punishable by one to ten years in prison, for allegedly conspiring to hurt people or damage property. The charges, based on an old law designed to stop coal miners from unionizing, were baseless: we had publicly pledged not to do either.

The felony charge was unexpected—and frightening. In the days that followed, I experienced mixed emotions. I was proud that I had participated in the action, taking risks to draw attention to a profound injustice. I was worried about the prospect of being entangled with West Virginia courts unlikely to be sympathetic to our cause.

We were fortunate to have an excellent lawyer, retained by the organizers, who had decades of experience fighting the coal companies that have despoiled West Virginia while leaving the state second only to Mississippi for the lowest life expectancy in the nation. In preparation for a preliminary hearing, he gathered statements from the defendants and the two volunteer legal observers, and subpoenaed body camera footage from the police. Seeing that there was no evidence supporting the charge, the prosecutor moved to cancel the preliminary hearing and did not pursue the charges.

Reflections on Civil Disobedience and Non-Violent Direct Action

So, what have I learned? Because most people don't expect older people like us to risk arrest, our participation in civil resistance can help to amplify key messages. As elders, we may be treated less harshly than younger people. Moreover, unlike young climate activists, we don't have to worry that an arrest record will hinder our careers.

Still, any action that challenges authority entails risk, and some actions are riskier than others. Actions at frontline fossil fuel sites, like MVP actions and the Coal Baron Blockade, are inherently riskier than protesting at the White House or in front of a bank. These are often in remote locations and local authorities may be inclined to file more serious charges. But these locations also hold symbolic power that may make the extra risk worthwhile.

It's important to assess each situation separately, carefully weighing the potential costs and benefits of various roles. If you are considering risking arrest, attend multiple civil disobedience trainings—many groups offer them—and for the first few times you join an action that potentially involves arrest, opt for a green or yellow role.

In the case of the two DC actions, the risks were modest and the benefits of the actions were eventually significant: the large, highly visible protests helped to energize the climate action movement nationwide, and Keystone XL was eventually canceled. (It fell to President Biden to finally pull the plug after he took office in 2021.)[62]

In the case of the MVP, the outcome is still unknown but there can be little doubt that the Yellow Finch tree sits, the "Old Hills & Old Folks Resist" blockade and other actions helped to delay construction and bring national attention to the fight. These actions also helped to build the movement: several boomers who met at these events went on to found Th!rdAct Virginia.

The Coal Baron Blockade garnered extensive, mostly favorable coverage in West Virginia media and a solid AP story that received national distribution. While we will never know for sure, it's possible that the bad publicity, together with many other pressures, led Manchin to strike a deal to pass the Inflation Reduction Act, the scaled down version of Build Back Better that is nonetheless the biggest climate investment package ever passed by the U.S. Congress.

Picking targets for climate-related civil resistance poses a challenge. For the Civil Rights movement, the targets were straightforward: schools, lunch counters and buses were segregated. When Black people showed up they exposed the injustice and desegregated these spaces. It was similar for many of the Vietnam War protests: by burning their draft cards, young men demonstrated their refusal to participate in the war and directly impeded the operations of the military draft.

For climate protests, the most relevant targets are coal mines, oil wells, powerplants and pipelines. Impeding such facilities sends a clear and powerful message. But they are often remote, making it hard for actions to generate publicity, and heavily guarded, making actions riskier. As we shall see in the next chapter, activists who physically interfere with fossil fuel infrastructure have been unjustly charged as terrorists, receiving heavy fines and long prison sentences.

Symbolic civil disobedience, like the Keystone XL protesters blocking of the "postcard window" in front of the White House, can be highly effective if the target of the protest—in that case

President Obama—is closely associated with the location. But other disruptive actions, such as blockading roads or mass transit—can backfire by alienating potential supporters.

In late 2022, climate protesters mounted a series of dramatic actions in half-a-dozen countries, blocking traffic, gluing themselves to famous artworks and, in one controversial instance in London, throwing tomato soup on a Van Gogh painting of sunflowers (covered by glass, the painting suffered no damage). The actions sparked a flurry of criticism to which the protesters replied: "We have tried everything else and we are running out of time. If you have a better idea, let's see you do it!" I tend to agree.

Assessing risks and effectiveness can be difficult ahead of an action. Plans are often fluid and take shape only in the final few days, or even the last hours. The only way to find out what's going to happen—and decide if you want to be part of it—is to join in the organizing and training sessions. As older people, we have lots of experience taking the measure of various groups and situations to decide if we want to be part of them. Use these skills.

If you are considering joining in civil disobedience for climate action, arrive early and watch and listen closely. Organizers are generally very careful not to push people into risking arrest, encouraging participants to assess their own situation and appetite for risk. Don't rush: for the first few actions you attend, opt for a green or yellow role. For every action that I have been part of, the people who risked arrest were a tiny proportion of the total.

Ultimately, however, if we are to effectively sound the alarm, we will need many thousands—even millions—of climate activist boomers who are prepared to risk arrest. Peter Kalmus, the NASA climate scientist who is an outspoken member of the Scientist Rebellion, said it well after his first arrest:

If you are concerned about climate change, the most liberating thing you can do is to risk arrest and possibly get arrested. You will be joining the ranks of people taking risks for other people. The feeling of solidarity with others is an incredible thing... This is so much bigger than all of us. We are not doing this for ourselves, we are doing this for our gorgeous planet and for each other.

Chapter 7: Should I Get Arrested?
Action Checklist

- Learn why civil disobedience is an American tradition
- Understand the three levels of risk: green, yellow, red
- Attend civil disobedience training sessions
- Support others who are prepared to risk arrest
- Ask yourself: why not me?

Who you callin' radical?

Successful social justice movements have radical flanks that are prepared to destroy property and even to use force in other ways. The climate movement urgently needs a bigger radical flank. How you can help, without getting hurt.

Chapter 8

Who you callin' radical?

Given the extreme urgency and profound, lasting impacts of the climate emergency—nothing less than destruction of the natural systems that support civilization as we know it—it is surprising that the climate movement has mostly colored within the lines, using conventional tactics up to civil disobedience and non-violent direct action, but no further.

Why is that? What other tactics are potentially available? Is limiting climate advocacy to the tactics described so far in this book a smart move that catalyzes action while building support? Or is eschewing other methods the equivalent of fighting with one hand tied behind our back? And, for the purposes of our inquiry, what role, if any, is there for boomers in supporting more radical actions?

Hypothetically, the range of possible actions goes well beyond civil disobedience. Kim Stanley Robinson's landmark climate novel, *Ministry for the Future*, explores many possibilities, including violence. In his harrowing opening chapter, set in the near future, a heatwave in India kills 20 million people in a week. A shadowy, tech-savvy group known as the Children of Kali arises from that disaster. Kali is the Hindu goddess of power, time and change, who destroys evil to protect the innocent. The Children of Kali do just that.

One brief chapter is written in the voice of an unnamed Children of Kali assassin who crawls through an air duct to drop onto the bed of a "guilty one"—presumably an oil company executive. "They killed us so we killed them," the chapter begins. Having stabbed the victim to death, the assassin clambers back through the air duct and makes their escape via drone. "Now to spread the headset photos, spread the story.

The guilty need to know; even in their locked compounds, in their beds asleep at night, the Children of Kali will descend on you and kill you. There is no hiding, there is no escape."

In another chapter, 60 passenger jets around the world mysteriously crash in a single day. "A disproportionate number of these flights had been private or business jets, and the commercial jets that had gone down were mostly occupied by business travelers," Robinson relates. About 7,000 civilians die that day in crashes caused by clouds of small drones that foul the aircraft engines. Jet travel pauses, then stops altogether after a second round of crashes the next month. Emissions-free blimps and dirigibles replace the planes. In other short chapters, drone torpedo attacks on container ships end global shipping; and drone darts infect cattle herds with mad cow disease, ending the beef industry, a major source of emissions.

Robinson's thriller is unusual in the burgeoning genre of climate-fiction ("cli-fi"): it shows humanity averting climate collapse. By the end of the 576-page novel, around year 2040, the fossil fuel era has ended and the world is pulling back from catastrophe. Political scientist Francis Fukuyama called the book "ludicrously unrealistic," for positing "the most optimistic possible political developments at every turn."[63] Yet even with optimistic political assumptions, Robinson evidently felt that a hopeful ending would only be credible with the Children of Kali meting out extreme violence to catalyze rapid systemic change.

Of course, a big advantage of fiction over other forms of discourse is that Robinson doesn't need to say whether he thinks such things *should* happen. Rather, he has imagined the world traveling a narrow path that just barely avoids planetary catastrophe, and violence is a necessary part of the journey.

Like many of Robinson's readers, I have complex and contradictory feelings about these brief vignettes. I am, of course, repulsed and terrified at the thought of such extreme violence. I can't imagine personally engaging in such acts. Yet, rationally, it's hard to imagine how, without some form of catalyzing violence, humanity can overcome the forces that are driving us toward a catastrophe in which millions and perhaps billions of people will perish. To be honest, there's a certain grim satisfaction in imagining that such rough justice could prevail—rough justice being better than no justice at all.

I recount these excerpts from Robinson's thriller to establish a continuum of tactics in which to consider the following two examples of ways to build a radical flank: hunger strikes and property destruction. Considered in relation to the actions discussed in the previous chapters, these tactics seem extreme. Considered in relation to Robinson's imagined Children of Kali, they seem moderate. They are also moderate, of course, compared to the extreme actions of the fossil fuel companies and their enablers, who are knowingly destroying the prospects for a livable planet to enrich themselves and their shareholders.

Before turning to tactics, however, I want to examine some strategic questions: What's a radical flank? Does the climate justice movement need one? If so, what role, if any, can boomers play in creating and sustaining one?

Why the Climate Movement Needs a Bigger Radical Flank

Andreas Malm, an associate professor of human ecology at Lund University in Sweden, is perhaps the best-known example of a small but growing number of activists and political theorists who argue that the climate movement cannot succeed without a radical flank. Title notwithstanding, his 2021 book *How to Blow Up a Pipeline* is not a manual but a closely reasoned argument, grounded in history, for the climate justice movement to develop a radical flank that includes property destruction in its repertoire of tactics.

Malm begins with a discussion on the reasons that mainstream environmental groups, and even smaller activist groups that focus on civil disobedience and non-violent direct action, have eschewed the use of force. Their reasoning, he argues, is grounded in what he calls "strategic pacificism"—a belief that peaceful civil disobedience is the most effective way to achieve climate action because it sounds the alarm and draws new supporters to the cause, whereas the use of force would provoke a fierce counter-reaction, alienating potential supporters.

He traces this widely held idea to the Chenoweth and Stephan study I discussed in the previous chapter, *Why Civil Resistance Works: The Strategic Logic of Nonviolent Conflict.* Unlike the many writers who cite the study approvingly, Malm takes Chenowith and Stepan to task for misreading history. Drawing on in-depth historical studies, he shows that the movements most often cited as examples of the success of non-violence all benefitted from a radical flank that was prepared to destroy property and sometimes use force in other ways.

For example, the Suffragists smashed windows and exploded a bomb along a royal parade route in London (nobody was killed in that or any other Suffragette action); Dr. King disavowed violence but the white establishment's fear of Black radicals—and recurring riots in American cities—made his demands seem moderate by comparison, helping to win passage of the 1964 Civil Rights Act and its 1968 enhancements; Mandela's support for non-violence was highly conditional, he described it as "a tactic that should be abandoned when it no longer worked."

Acknowledging these historical facts, Malm argues, would require that climate advocates who draw a line at non-violent resistance explain how the struggle for climate justice is an exception to the rule that struggles challenging a powerful *status quo* needs a radical flank to succeed. Those who insist that climate movement tactics go no further than peaceful means, he writes, would need to argue that while violence occurred in the struggle against male monopoly of the vote, against British and other colonial occupations, and against Jim Crow and apartheid, the "the struggle against fossil fuels is of a wholly different character and will succeed only on the condition of utter peacefulness."

While it is true that humanity has never faced a threat like this before, the differences between the struggle for climate justice and previous social justice movements point in the opposite direction: to the extreme difficulty of achieving change only through peaceful means.

Malm argues that ending the fossil fuel era comes closest to ending slavery, because both systems involve the unjust use of "productive forces" (e.g., fossil fuels and enslaved people) to create great wealth, power and privilege. In both cases, he says, the people who enjoy these huge benefits will not relinquish their so-called "property rights" without a fight. Moreover, just as the pre-Civil War U.S. economy was based on slavery, the entire U.S. economy today depends on fossil fuels.

Radical environmentalist Aric McBay makes similar points about the importance of a radical flank in *Full Spectrum Resistance*, a two-volume *magnum opus* that aims to debunk what he calls "the myth of pacifist persuasion." Citing a book by Lance Hill, he recounts an often over-looked aspect of Black Americans' struggle against racism: starting in 1964, Black people in the South organized self-defense groups known as the Deacons of Defense, with hundreds of active members and thousands of supporters. McBay quotes Hill: "The Deacons guarded marches, patrolled the Black community to ward off night raiders, engaged in shoot-outs with Klansmen, and even defied local police in armed confrontations."

Rather than tracing the mainstream climate movement's aversion to use of force to a single political science study, as Malm does, McBay sees it as grounded in the interests of what he calls, drawing on the work of journalist Chris Hedges, the "liberal class," a group that includes me and many of my readers: "parts of the press, labor unions, and universities, as well as liberal religious institutions and elected officials (like the Democrats)." The problem with liberals like us, McBay writes, is that we are unwilling to risk our comfort and privilege to achieve justice.

"I would have a lot more respect for liberals in general if they were more honest. Imagine a liberal said to me: 'I want to have social change and a livable planet, but I'm actually very comfortable, and I'm too scared to risk that.'" Ouch! For me, these words cut painfully close to the bone.

Worse, McBay argues, we liberals have recast history with ourselves as the heroes. Liberals, he writes, "have claimed credit for victories won through radical action, like the end of segregation, or women's right to vote, or the eight-hour day. When social struggles in history have achieved victory using a diversity of tactics, liberals have expunged the 'dirty' tactics from their version of the story until they have a narrative that is tidy, sanitized, and wrong."

The problem with such a narrative, McBay and Malm would agree, is that it obscures the critical need for a diversity of tactics, especially the crucial role of a radical flank. McBay says that successful movements are like healthy ecologies: highly diverse. "A group should be consistent internally, but that doesn't mean it should be the same as all other groups in a movement," he writes. "It makes perfect sense for a movement to include both large, moderate organizations and clandestine militants." Although he doesn't carry the ecology analogy this far, I suggest that we think of movement militants as being like apex predators: relatively few, fierce, scary, and critical to the functioning of the entire system.

Shifting the Overton Window

In a well-functioning culture of resistance, with a full diversity of tactics, actions on the fringe of acceptability benefit a movement in many ways, McBay writes. "Militant action makes a moderate position (and the possibility of compromise) much more appealing to those in power, and it makes formerly risky action appear more moderate."

This dynamic is part of a phenomenon known as the "Overton Window," named for Joseph Overton, a libertarian boomer who worked for a small right-wing think tank in Michigan. To raise money for the tank, he designed a brochure showing how public perceptions of potential policies can be shifted along a continuum: from unthinkable, to radical, acceptable, sensible, and popular, culminating eventually in actual public policy. Think tanks, Overton said, played a crucial role, by introducing ideas that were previously unthinkable, then relying on allies, such as sympathetic media, to move them along the continuum until they become policy.

The American right has enthusiastically embraced the concept of the Overton Window, sometimes also known as the "window of discourse." McBay and others note how extreme right wing policy proposals and so-called "conspiracy theories" (which I prefer to call crackpot lies) originate on far-right Internet chat boards and are picked up by right-wing media outlets that constantly push the boundaries of acceptable discourse until ideas that were once unthinkable become mainstream.

During the 1960s and 1970s, left radicals also shifted what we now call the Overton Window. Dr. King considered the early Freedom Riders—racially mixed groups of young people who rode public buses into the segregated South—as too radical. "He spent the entire night before the first group departed trying to convince the riders to call it off," McBay writes, citing historians of the movement. But the riders went ahead, broadening the spectrum of acceptable protest. Similarly, armed groups like the Deacons for Defense and later the Black Panthers made the demands of non-violent leaders like Dr. King seem more acceptable, opening the way for compromise.

To be effective, a radical flank must walk a precariously narrow line, Malm writes. On one hand, it must go beyond what the existing movement endorses—and be prepared to be loudly denounced. "Prospective militants should expect and even hope for condemnation by the mainstream, without which the two would become indistinguishable and the effect would be lost," he writes.

On the other hand, the risk of backlash from society must not be underestimated "Extremism can make a movement look so distasteful as to deny it all influence," he writes. "There is no lack of examples of movements shooting themselves in the foot. Because of the magnitude of the stakes in the climate crisis, negative effects could be unusually ruinous here." Nonetheless, he writes, public receptivity to the actions of a radical flank, even including sabotage of fossil fuel infrastructure, is likely to change over time, as the extreme damages of climate disruption become more evident.

"If fossil fuels continue to be combusted and temperatures to climb, physical attacks on the sources... of the calamities should resonate with more and more people," Malm writes. "The problem is that to blow up a pipeline in a six-degree world would be to act a little late. Should we wait for approval from a near consensus? A majority? A big minority? The task of climate activists cannot be to take the existing consciousness as a given, but rather to stretch it."

A note of clarification: shifting the Overton Window applies to policies—as in Overton's original think tank-focused formulation—as well as to tactics. For example, members of the youth-led Sunrise movement study the Overton Window in their training sessions and used it effectively in putting forward an ambitious climate, jobs and justice proposal—the Green New Deal. Progressive Democrats, like Sen. Bernie Sanders and Rep. Alexandria Ocasio-Cortez championed it, creating space for moderates, like President Biden and House Speaker Nancy

Pelosi, to put forward a scaled-back version, Build Back Better. The Inflation Reduction Act that eventually passed fell well-short of the original Sunrise proposal but arguably would never have happed had it not been for the Green New Deal.

While policy proposals and activist tactics are distinct, they are closely linked. Ideas that may appear to be at the outer edge of the policy debate, like "no new fossil fuel infrastructure," are championed by groups that use radical tactics, such as the #JustStopOil activists who threw soup on the Van Gogh painting, and the #DeclareEmergency activists who blocked the Capital Beltway around Washington, DC. An effective radical flank uses both policy proposals and protest tactics to broaden the terms of the debate and shift the consensus about what constitutes a reasonable course of action.

How Boomers Can Help Build the Radical Flank

Having understood the crucial importance to the climate movement of having a radical flank, what can we do about it? There is a range of options. For those who are willing and able, joining directly in such actions can unlock new energies and power to address the climate emergency. Of course, radical flank actions are very risky and should only be undertaken with full recognition of the potential, serious consequences.

Fortunately, supporting a radical flank doesn't require getting arrested, much less risking life, limb or liberty. Even an environmental radical like McBay, as disdainful of liberals as he is, sees roles that are a better fit for most climate boomers:

> *Not everyone is able or willing to take the greatest personal risk; not everyone will blockade a pipeline or take over a city council meeting. That direct action is the job of only a small percentage of people at any given time. The job of the majority of people in a culture of resistance is to support those few. They need support*

morally, through vocal support for militancy and advocacy for resistance. They also need material support through food, fundraising, childcare, prisoner support, and all the rest. A wide base of material support is needed to win any conflict.

Speaking out in support of a radical flank, preparing meals, donating or raising money, helping take care of people's kids, and visiting and corresponding with prisoners. These sound to me like things that those of us who are old enough to be grandparents could be particularly good at. I have begun doing this in small ways, contributing to bail funds and writing letters to people jailed for radical climate actions. I hope that many more of us will do the same.

So, what does radical flank action look like? And how might we support it? The rest of this chapter considers two examples: hunger strikes and the use of force, including vandalism and sabotage.

A Short History of Hunger Strikes

Refusing to eat to protest injustice has a long history. As the climate emergency worsens, climate-related hunger strikes are becoming more frequent, yet they are still relatively rare given the urgency and severity of the crisis. Knowing the history of such actions can inform our consideration of this tactic for addressing the climate emergency.

Hunger strikes should not be undertaken lightly. Long-term refusal of food affects most organs and systems in the human body, causing muscle weakness, vulnerability to infections, psychological problems, and, eventually, organ failure. If a protester is healthy before starting a hunger strike, and continues to receive fluids, they risk dying from malnutrition after six to eight weeks. But a hunger striker may die much sooner—after three weeks if they're seriously ill. If refusing fluids, death can come much faster, after one week.

In some instances, the willingness to take on these extreme risks can force the powerful to change their ways. According to legend, a hunger strike can even challenge the will of God.

A ninth-century account of the life of St. Patrick tells how he climbed a mountain in Ireland and fasted for some 50 days while negotiating with God about how many souls he could save. St. Patrick announces he will fast: "until I am dead, or until all my requests are granted." Each time God agrees, St. Patrick moves the goal posts. Save seven souls every Saturday till Doomsday? OK, says God. Make it twelve, says St. Patrick. OK, says God. More demands and concessions are exchanged. In the end, God grants St. Patrick the power to judge all Irish souls on Doomsday.

The world's best-known hunger striker is Mahatma Gandhi, who undertook 18 fasts. The goals differed, from objecting to his detention by the British, to supporting striking mill workers, to ending violence between Hindus and Muslims, to supporting the rights of the low-cast "untouchables." Most of Gandhi's fasts lasted a week or less. Two lasted 21 days each, including one for Hindu-Muslim unity and another to protest his detention by the British (the British ignored him and nothing changed).

The most famous hunger strikers in the United States were Suffragettes who refused to eat to protest the poor conditions they encountered in prison. In November 1917, Lucy Burns and other Suffragettes on a hunger strike at a workhouse south of Washington, DC, were brutally force-fed in what became known as the "Night of Terror." The resulting uproar helped build momentum for passage of the 19th Amendment guaranteeing women the right to vote.

In 1972, Mexican American labor organizer Cesar Chavez staged a 24-day hunger strike opposing a new Arizona law that restricted farmworkers' right to organize. Increasingly emaciated, Chavez regularly appeared at Mass, where he was joined by Dr. King's widow, Coretta Scott King, and Democratic presidential nominee George McGovern. Chavez referred to the strike as "a fast of sacrifice," saying that his suffering represented the daily suffering of farm workers.

The most extreme instance of hunger strikes in modern times occurred in 1981 at the culmination of a five-year protest by Irish republican prisoners in Northern Ireland that became a showdown with Prime Minister Margaret Thatcher. One hunger striker, Bobby Sands, was elected to parliament during the strike. Thatcher refused their demands and ten prisoners, including Sands, starved themselves to death. The strike radicalized Irish politics and was a driving force in the Irish republican political arm, Sinn Féin, becoming a mainstream political party.

Hunger Strikes for Climate Action

The past two years have seen a series of climate hunger strikes, including by a group of young people in Berlin, single hunger strikers in Bern and London, and five young Americans who staged a two-week long hunger strike in front of the White House.

In September 2021 young German climate activists held a hunger strike outside the Reichstag demanding that candidates for chancellor meet with them publicly before the election. Olaf Scholz, the Social Democratic Party front-runner, agreed to meet with them if he was elected, a promise he kept. During the meeting he rejected their accusation that German politicians were not taking the issue seriously.

A Swiss hunger striker, Guillermo Fernandez, 47, fasted for 39 days outside the Swiss parliament, demanding it host a climate science briefing. An IT programmer and father of three, he told reporters he started thinking about a hunger strike on his younger daughter's birthday. Swiss scientists issued a letter backing his demand and held one-day fasts in support. He ended his fast in December 2021 when the Swiss Academy of

Natural Sciences announced that members would brief Swiss parliamentarians on climate change.

In March 2022, Angus Rose, a 52-year-old British electronics engineer, began a hunger strike in demanding that the government's chief scientific advisor publicly brief Parliament on the climate emergency. Prime Minister Boris Johnson had said that a similar confidential briefing had awakened him to the magnitude of the threat. Rose ended his hunger strike after losing 37 pounds in 37 days when Parliament agreed to host the briefing. None of the Conservative Party MPs then in the running to succeed Johnson attended.

In the United States, five young people belonging to the Sunrise movement began a hunger strike outside the White House in October 2021, ahead of the U.N. Climate Conference in Glasgow, Scotland. They demanded that President Biden do everything in his power to win passage of Build Back Better, the administration's scaled-back version of the Green New Deal. During the election, hundreds of Sunrise volunteers had knocked on doors for the Democratic ticket, helping to secure President Biden's victory.

"I'm waiting for Joe Biden to start fighting for the people who elected him," said Kidus Girma, who at 26 years old was the eldest of the group. During their strike, the group met online with White House National Climate Adviser Gina McCarthy and Special Presidential Envoy for Climate John Kerry. Hunger striker Paul Campion, 24, told CNN that Kerry had assured him that climate negotiations in Congress were in a good place. "He kept saying there were complications in the Senate," Campion said. "I was pretty furious and pretty disappointed."

During their hunger strike, more than 250 people in the United States and at the U.N. climate conference in Glasgow joined a 24-hour All Saints Day solidarity fast. "Today, we follow these young adults," said Anna Robertson, director of youth mobilization for the Catholic Climate Covenant, opening an online support vigil. "To keep vigil is to wait, to wait firmly, to wait solemnly," she said. "We are waiting for the end of unnecessary suffering. We are waiting for the end of the cycles of domination and oppression."

The young people ended their hunger strike after 14 days, when President Biden reaffirmed an earlier pledge to cut fossil fuel emissions in the U.S. in half by 2030.

Hunger Strike Lessons for Boomers

What lessons can we draw from these experiences? One is to formulate demands that are worth fasting for yet conceivably within reach. The Swiss and UK hunger strikers seem to have had this in mind when they demanded that their parliaments be briefed by climate scientists. The parliaments agreed, the hunger strikes ended, and the briefings were held, but the politicians who most needed to hear the information refused to attend. Similarly, the German hunger strikers achieved their goal of meeting with the newly elected chancellor but little came of the meeting.

The Sunrise hunger strikers in the United States set a high but ambiguous bar—greater efforts by President Biden to persuade the Congress to pass his signature Build Back Better legislative package. They ended their strike when the president repeated a previous pledge in a major global forum.

Seen narrowly, the German, Swiss and UK hunger strikers could claim to have achieved their goals, while the Sunrise hunger strikers in the U.S. did not. Seen more broadly, though, each of these actions helped to draw attention to the issue and sound the alarm, possibly strengthening a public consensus for action.

Another lesson is to pick the target carefully. Legend has it that St. Patrick successfully bargained with God. Thatcher proved a tougher nut to crack, refusing to negotiate with the

Irish republicans, even as ten of them starved themselves to death.

Should climate boomers consider hunger strikes? On one hand, many of us have health conditions that could make a hunger strike riskier for us than for younger, healthier individuals. On the other, as with civil disobedience, we have less to lose than young people. After all, most of our lives are already behind us. While we may be less resilient, a hunger strike that injures the health of an older person is less of a tragedy than a similarly heavy toll imposed on a young person with most of their life still ahead of them.

More importantly, our position as elders could make a boomer hunger strike a powerful tactic. Imagine, if you will, a mass, public hunger strike by scores of us simultaneously in the capitals of each of our 50 states. Done right, such an action could attract a great deal of attention, creating more pressure for corporate and government decisionmakers to act. The challenge is to identify a concrete demand that would be worth a mass hunger strike yet conceivably within reach.

A less dramatic—and less dangerous—alternative could be public fasting vigils, where the duration of the fast is announced at the start. Fasting as a form of penance has roots in many religious traditions and could be explored as a tactic for the climate movement.

Vandalism as Direct Action: The Dakota Access Pipeline

This is the story of two brave women, Jessica Reznicek and Ruby Montoya, who were inspired by 1960s anti-war activists to vandalize the Dakota Access Pipeline. I believe that Reznicek and Montoya are heroes. Yet even in climate action circles, many people do not know their names. Learning Reznicek and Montoya's story, sharing it with others, and joining public appeals for justice in their cases are examples of how climate boomers can build the radical flank without even leaving home.

Since Reznicek and Montoya's action, the Dakota Access Pipeline, also known as DAPL, has been completed and now carries crude oil from the Bakken oil fields in western North Dakota, through South Dakota and Iowa to southern Illinois, crossing land sacred to Native Americans and the Missouri and the Mississippi rivers. Like other big fossil fuel pipelines, it endangers land, water and people while extending the fossil fuel era, intensifying the risk of a planetary catastrophe.

During construction, thousands of people traveled to North Dakota in 2015 and 2016 to support Native Americans and other local people trying to block the pipeline, as lawsuits made their way through the courts. After a stand-off of many months, the last of the activists were removed by court order in early 2017. Energy Transfer Partners, the Texas-based developer behind the pipeline, began final steps to complete the project.

Against this background, Reznicek and Montoya called a press conference in July 2017 and announced that they had vandalized the pipeline. They said that they began their action on Election Day 2016—the day that Trump won the electoral vote despite losing the popular vote by a record 3 million ballots—by setting fire to heavy equipment on the pipeline route. Later they pierced an empty portion of the pipeline using oxyacetylene torches, delaying completion for weeks.

They stressed that their actions never endangered people. "Both myself and Jessica acted safely," Montoya said. "The fires that we started, those were very contained. These were empty construction sites. They had already desecrated the trees by clear-cutting. There was a lot of prairie land and that was all cleared. There were no workers on site at any time."

They said that they were inspired by the teachings of the Catholic Worker Movement. Founded in the 1930s by American journalist and political activist Dorothy Day, who is now being considered for sainthood, and Peter Maurin, a French Catholic activist and theologian, the movement continues to operate through some 240 autonomous communities, known as houses, which provide social services around the world. The movement is dedicated to non-violence and opposes war and the unequal distribution of wealth.

Reznicek and Montoya had lived in a Catholic Worker Movement house in Des Moines, Iowa, named after Philip Berrigan who, together with his brother Daniel, a Jesuit priest, became famous in the 1960s for daring actions opposing the Vietnam War, such as breaking into government offices to destroy draft records by pouring blood on them, and throwing the records outside and setting them on fire. Philip Berrigan spent 11 years in prison. Reznicek and Montoya said that they saw themselves as continuing in the tradition of the Berrigan brothers.

"Our conclusion is that the system is broken and it is up to us as individuals to take peaceful action and remedy it," Montoya later told Michael J. O'Loughlin, a reporter for the Jesuit publication *America.* Reznicek said that she had attended Catholic schools as a child and that her social justice work was rooted in Catholic tradition and scripture. Her actions, she said, were an "opportunity to walk in a way that has integrity with the scriptures and with what Christ has taught."

The two were charged with nine federal criminal charges each. Reznicek was initially sentenced to four years, but the judge labeled her a "domestic terrorist"—despite the fact that she and Montoya never threatened or endangered people—and doubled her sentence to eight years. Montoya initially pleaded guilty, then applied to withdraw her guilty plea, a request that was denied. In June 2022, a judge denied Reznicek's appeal against the so-called terrorism enhancement of her sentence.

Her appeals exhausted, Reznicek was reflective in a video interview before reporting to prison. "I started settling into the idea that I was going to dedicate my life to something," she said. "I didn't know that would look like prison *per se,* but to dedicate your life to something means to give it your all, without holding back that sacrificial love that lives in my heart."

Julia Steinberger, a lead author with the UN's Intergovernmental Panel on Climate Change (IPCC), is among the many scientists who have spoken out in support of Reznicek. "As the climate and ecological crises escalate, so too will our activism," she wrote. "And so will the repression by governments corrupted by fossil fuel interests. Stand with Jessica Reznicek."

In the case of Reznicek, supporting the radical flank is simple. Start by signing the petition at SupportJessicaReznicek.com demanding that President Biden commute her sentence or pardon her, then tell your friends and follow and share @FreeJessRezz on Twitter, Instagram and Facebook. By late 2022 scores of organizations—including many committed to non-violence—had signed the petition, along with more than 16,000 individuals.[64]

Reznicek and Montoya weren't the first to physically interfere with fossil fuel infrastructure. In October 2016, five people who came to be known as the Valve Turners simultaneously closed emergency shut-off valves on oil pipelines in four American states. All five had planned to use the "necessity defense," arguing that their actions were necessary to prevent a greater harm. Judges permitted this in only two of the cases. Only one of the Valve Turners, Michael Foster, who was 53 at the time, received a prison sentence, serving one year with two years suspended.

As IPCC scientist Steinberger notes, we can expect to see more such actions as the severity of the climate crisis—and the refusal of the fossil fuel industry to change their planet-destroying business model—become increasingly evident. Climate boomers who understand the need for a radical flank can be on the lookout for these actions and ready to offer support when they happen.

Eco-Sabotage: British Columbia's Coastal GasLink Pipeline

Not everybody who blocks fossil fuel infrastructure feels compelled to step forward and announce what they have done. Malm's introduction to an April 2022 collection of essays responding to *How to Blow Up a Pipeline* begins with a harrowing description of a February 2022 midnight attack on a British Columbia methane pipeline that makes Reznicek and Montoya's action look like a stroll in the park.

According to Malm, 20 masked attackers seeking access to the construction site emerged from the dark woods, surrounded the truck of a security guard, and demanded that he open the gate:

> *One activist started cutting the gate with a cordless power tool, while others swung axes against the side of the truck. Flare guns, spray paint and strobe lights were also deployed, until all the security forces, overwhelmed by the sudden attack, fled their positions. The site... was now in the hands of the militants. They entered the driving seats of bulldozers and trucks and used them to smash other machinery. Generators, heavy equipment and modular trailers were hacked to pieces, bulldozers overturned, trucks shattered. Video surveillance systems were methodically gutted. When the police arrived, they were held back by barricades made of trees, spikes, fires and a yellow school bus; after clearing the way into the site, they found an empty scene of wreckage. No suspects remained.*

The pipeline company, a subsidiary of Calgary-based TC Energy called Coastal GasLink, reported millions of dollars in damage. Construction stopped on the 670-km pipeline, being built to transport LNG to the coast for export to Asia. A member of the provincial assembly worried that the action "could create a chill," discouraging investors from backing fossil fuel projects in the province. As Malm notes, that was precisely the point of the action.

Malm's account, which matches press reports at the time, recalls similar acts by EarthFirst!, a radical environmental group founded in 1980 in the southwestern United States. Inspired in part by Edward Abby's classic environmental novel, *The Monkey Wrench Gang* (1975), EarthFirst! engaged in direct action to prevent logging, dam construction, and other construction that would despoil wild places. During a 1985 action to stop logging in western Oregon, EarthFirst! pioneered the first tree sit, a technique now emulated in attempts to block new fossil-fuel infrastructure, including the Mountain Valley Pipeline.

Given the urgency of the climate threat, why aren't we seeing more such actions? Arnold Schroder, an EarthFirst! veteran, muses about the organization's strengths and failings in his fascinating if rambling podcast, *Fight Like an Animal*. "As grief and terror about the ecological crisis intensifies, it seems increasingly curious that for many years a radical environmental movement—based on a deep sense of connection with, and rage on behalf of, all life on earth—existed, but is now largely silent," he writes in the introduction to one episode.[65]

Part of the explanation, he says, is "the baffling complexity of fighting for a survivable climate rather than a specific place." In my view, the environmental justice movement, with its opposition to site-specific fossil fuel infrastructure grounded in concerns about the impact of pollution on frontline communities, is one way to square that circle: direct action to stop coal mines, fracking, pipelines and petrochemical plants can be simultaneously local *and* global, an attempt to stop local damage while also preventing increased emissions.

Reflections on Property Destruction

I find stories like the eco-sabotage of the Coastal GasLink Pipeline at once frightening and elating. Frightening, because things could so easily go wrong. Somebody could get hurt—or killed. And while I admire those willing to risk multi-year prison sentences for modest acts of anti-fossil fuel vandalism, I am personally unwilling to spend my final years in prison without knowing that my act would make a significant difference. But I also find these stories elating, because the stakes are so incredibly high that *not* resorting to force—when history suggests the climate movement is unlikely to prevail without it—seems to me to be morally indefensible.

Such stories about the use of force shift the Overton Window for two audiences. For activists, it reminds us that other forms of direct action—including non-violent direct action—are much less risky by comparison. I'm not willing to risk years in prison for cutting pinholes in an empty stretch of pipeline, or to wield an axe in a midnight attack on a pipeline construction site. But if I share the goals of those who are willing to take such risks—and I do—then perhaps I should be willing to risk arrest and spend several days in jail for an act of peaceful civil disobedience.

For corporate decision-makers and mainstream politicians, stories about the use of force shift the Overton Window in the same manner that the Deacons of Defense and the Black Panthers shifted white politicians' views of Dr. King and his demands. As the use of force to oppose fossil fuel infrastructure becomes more widespread—as seems likely—it may cause the rich and powerful to begin to heed the demands of peaceful climate activists engaged in non-violent civil disobedience, who up until now have appeared to be a radical fringe.

Climate boomers can boost the impact of radical actions with both audiences by keeping an eye out for news reports about the next such action by learning and telling the stories of the brave people who undertake these acts, and by providing financial, moral and public support like that being offered to Jessica Reznicek.

Summing Up, Looking Ahead

I imagine that this has been a hard chapter to read. It's certainly been challenging to write. Recognizing the extreme urgency of action, the profound social and economic transformations needed to avoid catastrophe, and the immense power of the vested interests blocking change, I explored the need for a radical flank. I have shown how radical actions can shift the Overton Window, so that ideas that were once unthinkable become mainstream and eventually move into the realm of policy.

I then considered two radical tactics, hunger strikes and the destruction of property. I showed how these tactics have been deployed historically to overcome injustices that, while horrific, are limited and transitory compared to the planet-wide, eons-long existential threat of rapid global heating. I noted that a handful of people are already resorting to these tactics out of a sense of desperation, and that their numbers will grow as the impacts of climate change become more frequent, severe and obvious.

For each of these actions I considered the potential roles of climate boomers.

For hunger strikes, the personal risk is potentially immense, including damage to health and even risk of death. Yet many people throughout history have undertaken hunger strikes to call attention to injustices that were more transitory than climate change. Within carefully thought-through parameters—including formulating demands that are worthwhile but attainable and a decision-maker target who is potentially persuadable—boomer hunger strikes should be an arrow in our quiver.

For destruction of property to defend the planet—from largely symbolic vandalism to outright sabotage—the personal risks are extremely high. Older people like us, lacking the strength and agility of younger generations, are often ill-suited to the physical demands. We can, however, help to build the radical flank by speaking out in support of those who take these risks, raising funds, and providing other forms of support.

Chapter 8: Who You Callin' Radical?

Action Checklist

- Learn why the climate movement needs a bigger radical flank
- Tell the stories of those who take these risks
- Donate or raise money for bail funds and to support radical groups
- Write letters to climate activists in prison or jail
- Offer food, lodging, help care for people's kids

Just do it!

You're never too old
to help save the planet.
By lending a hand, you
will enrich your life and
leave a positive legacy
that will endure for
generations to come.

Chapter 9

Just do it!

The growing global movement to save a livable planet can only succeed if millions of American boomers lend a hand. There are many ways to help. By getting involved now, you can leave a positive legacy that will endure for generations to come.

We have seen how high the stakes are and why our generation has a moral obligation to act. We are uniquely responsible for the climate emergency because we ran the country when climate change went from a manageable problem to an existential crisis. And because the United States is the largest historical emitter and the world's most powerful nation, the lack of U.S. leadership is having global repercussions.

We have also seen that our generation still has immense power. We control most of the country's wealth, we vote in large numbers, and we are rich in knowledge, skills and connections. By connecting our power with the ideals of our youth, we can play a key role in saving the planet for our children, grandchildren and future generations, in the United States and around the world. Indeed, unless a significant number of us organize to demand action, future generations will be condemned to struggle for mere survival on a hostile planet largely of our making.

These are the messages of Part I of this book. Part 2 is the core: a boomer's guide to the many ways that you can engage, from actions that you can do on your own, such as shifting your diet and installing rooftop solar, to working with volunteer and faith-based climate advocacy groups, to participating in or supporting civil disobedience and non-violent direct action. The penultimate chapter explains why the climate movement urgently needs a bigger radical flank, what this is, and how you can help to nurture one without putting yourself at risk.

This final chapter briefly reviews the options for action and offers tips and tools for deepening and expanding your participation in the global movement to save the planet. The title is borrowed from one of the most successful advertising slogans of our time: just do it!

Make the Most of Personal Actions

Many of the volunteers I interviewed for this book began their journey towards climate activism by taking one or more of the six personal actions described in Chapter 4: 1) stop wasting food and eat less meat; 2) drive less: walk, bike & take public transport more; 3) move your money; 4) upgrade your car to an EV; 5) install rooftop solar; and 6) fly less—or not at all.

You can begin the first and second actions right away, with gradual changes to what you eat and how you get around. To make the most of these and other actions—and to increase the likelihood that you will follow through on your good intentions—discuss your plans with your family and friends. Tell them why you plan to make these changes and invite them to join you.

Action three, moving your money so you stop propping up the fossil fuel industry, takes time and planning. If you use a commercial bank, check its standing in the *Banking on Climate Chaos* report. If yours is among the Dirty Dozen banks financing the fossil fuel industry, sign the Th!rdAct pledge to move your money and start investigating your options. To purge fossil fuels from your investment portfolio, check the guidance from As You Sow, or consult a Certified Financial Advisor, ideally one who is also a Chartered SRI Counselor (CSRIC). THIS offers checklists and free coaching sessions for groups of five or more. Do you have four friends who would join you? Ask around. You may be surprised!

Actions four and five, buying an EV and installing rooftop solar, involve upfront expenditure but tax incentives and new financing approaches can reduce the costs. If you are considering these steps, begin researching and planning both now, since each takes a bit of time. Check what incentives are available through the 2022 Inflation Reduction Act (IRA). Rewiring America offers a calculator that shows what incentives are available based on your income and location.

Action six, fly less or not at all, is the most difficult for many of us, including me. Flying results in more emissions than anything else we do and voluntary offsets are unlikely to deliver as promised. I suggest an alternative: if you must fly, make an (additional) donation to a highly-effective climate advocacy organization equal to the damage caused by your trip using the EPA's estimate of the social cost of carbon: $190 per ton. This is the amount of carbon emitted by one person flying economy coast-to-coast. The high cost may cause you to think twice next time you consider air travel, which is part of the point. See Chapter 4 for details!

Get Connected to Increase Your Impact

Chapter 5 showed how you can increase your impact, starting with a quick self-assessment of your strengths and networks. If you haven't done so already, take the free online StandOut Strengths Assessment. Are there strengths you want to use in your new role as a climate activist? Now reflect on your networks. What groups can you tap to find allies and amplify your impact?

Next, use national organizations to find local connections. I list seven that are effective in welcoming new volunteers: Chesapeake Climate Action Network (CCAN), Climate Reality Project, Elders Climate Action, Moms' Clean Air Force, Sierra Club, THIS Is What We Did, and 350.org. Shop around and see which is good fit, looking for one that matches your interests and values and has an active group of volunteers near you. Ask a friend to join you in attending a local meeting or showing up at a rally.

Besides these magnificent seven, I highlight two organizations that stand out for their reach and ambition. The first, Th!rdAct, organizes "experienced Americans" to become climate activists and protect democracy. Inspired by the activism of the 1960s, its members use public demonstrations and other tactics to press big banks to stop investing in fossil fuels. The second, Citizen's Climate Lobby (CCL), has a large network of volunteers who use a low-key, non-confrontational approach to create a bipartisan consensus for a specific policy solution: carbon fee and dividend. I explain why I believe that this policy, unpopular with many progressive climate groups, may nonetheless hold great promise.

The rest of Chapter 5 discusses three ways you can use your connections. First, use your status as a college or university alum: support student-led campaigns to divest from fossil fuels and challenge conflict-of-interest-ridden relationships with fossil fuel interests. Second, use your status as a retiree: join fellow boomers in demanding that your pension fund clean up its act. Finally, look for opportunities to act through your professional ties. I name three professions: healthcare, law, and advertising and public relations.

There are many paths to climate action and many ways to be an activist. Some people retire early to become full-time activists. Others take a break after retirement and are drawn to activism as they learn more about the climate emergency. Others, like me, were already active and step up their participation after retirement. Still others join the movement after a major life transition, such as losing a loved one or moving to a new community. Some volunteer full-time while others devote a few hours a week and are also making a difference.

Still seeking your path? Perhaps religion is the answer!

Tap the Power of Organized Religion

I devote a chapter to faith-based action because I believe it's an overlooked source of power. Chapter 6 explains that religion can be a powerful force for saving the planet because politicians understand that people of faith are motivated by moral convictions grounded in religious teachings. How to unleash this power? Boomers, the majority of the "people in the pews," can participate in faith-based advocacy and encourage our religious leaders to use the power of the pulpit to speak out more forcefully.

Participating in faith-based climate advocacy also offers us a personal benefit: community. As we shift into retirement, we tend to lose touch with our workplaces and former colleagues. Whether you are already connected to a religious community, looking to reconnect with the religion of your childhood, or perhaps on a spiritual quest, exploring faith-based climate advocacy can open the door to new friendships with people who share your desire to pass a livable planet to future generations.

How to begin? Find a find a local group, within your own tradition or interfaith, which meets regularly and will call you to action. Second, find the climate-relevant language and theology that is authentic to your tradition and use that to ask your faith leaders to do more.

I start the chapter by describing the central role that Native Americans are playing in leading and inspiring the climate action movement, drawing on their sacred teachings and traditions. I then introduce the climate-relevant teachings and faith-based advocacy of six religious groups: Catholics, evangelical Protestants, ecumenical Protestants, Black Churches, Muslims, and Jews. Each tradition has theological underpinnings and statements from top leaders urging action.

But these often aren't enough to persuade community-level religious leaders to make climate action a part of their ministry. Clergy set their priorities based on the guidance from senior religious authorities and what they hear from their followers. They are probably worried about climate change themselves and they may be waiting to be nudged to do and say more. Be the nudge! If you belong to a religious organization, meet with a clergy person to share your deep concern and ask how the congregation can do more.

Are you inspired to explore these paths? Now would be a good time to re-visit Chapter 6, perhaps re-reading the section that connects most closely with your own faith background. Then follow the tips at the end of the chapter on ways to get involved. I guarantee that your inquiries will be warmly received.

Support Non-Violent Civil Disobedience and Direct Action

Chapter 7 is titled "Should I get arrested?" The answer depends on you. But given how high the stakes are, I believe we should all at least be asking ourselves this question. We know from the continued rapid rise in emissions and the growing number of climate-related disasters that unless societies and economies are rapidly transformed, humanity faces a climate catastrophe that will last for millions of years. Surveys show that a growing number of Americans—including millions of boomers—say that they would be willing to risk arrest to stop government or corporate actions that make climate change worse.

I begin with a historical perspective, showing that risking arrest to shine light on injustice is an American tradition dating back to Henry David Thoreau's famous 1847 essay "On the Duty of Civil Disobedience." Thoreau's essay deeply influenced the world's most famous leaders of struggles for social justice, including Mahatma Gandhi and Dr. Martin Luther King. Civil disobedience was also central to opposition to the Vietnam War.

The 1960s-era civil disobedience inspired the climate movement, including mass arrests outside the White House to stop the Keystone XL Pipeline. Since then, climate activists have engaged in hundreds of acts of non-violent civil disobedience across the country. Their actions have helped to raise awareness, delaying and sometimes halting fossil fuel projects that would have damaged local environments and locked in higher emissions for decades.

I describe several instances where boomers have participated in civil disobedience, including my own and others' experiences in being arrested. I conclude that boomer civil disobedience is among the most effective tools we have for drawing attention to the climate emergency. We can make use of our status as elders: police and judges may treat us less harshly and, with our working lives mostly behind us, an arrest record won't hurt our employment prospects.

Assessing risks and effectiveness in advance of an action can be difficult. Plans are often fluid and emerge only in the final few days, or even the last hours. The only way to find out what's going to happen—and decide if you want to be part of it—is to participate in the organizing and training sessions. Arrive early and watch and listen closely. Don't rush: for the first few actions you attend, opt for a supporting role that does not involve risking arrest.

Ultimately, however, if our generation is to help sound the alarm, we will need many thousands—even millions—of climate boomers to risk arrest. I urge you to ponder carefully the question that begins the chapter: should I get arrested?

Help Build the Radical Flank

The penultimate chapter delves into territory that is rarely explored outside of the world of fiction and the writings of radical environmentalists such as the two that I cite, Andreas Malm, author of *How to Blow Up a Pipeline*, and Aric McBay, author of *Full Spectrum Resistance*. Drawing on their writings and the work of the historians they cite, I argue that major social justice movements need radical flanks—groups that take positions and adopt tactics outside the mainstream.

History shows that radical flanks have repeatedly opened the way for progress by causing those with power to accede to more moderate demands. I explain how radical flanks can shift the Overton Window, also known as the window of discourse, opening space for the acceptance of less radical proposals that were previously seen as unreasonable or impossible.

Along the way, I shift the Overton Window within the chapter itself, telling stories from Kim Stanley Robinson's cli-fi thriller, *The Ministry for the Future*, in which a shadowy group of high-tech militants take revenge for a heat wave in India that killed 20 million people in a week. Their violence helps to bring a rapid end to the fossil fuel era, narrowly averting a planetary catastrophe. I tell these frightening stories to open the way for consideration of two tactics that are beginning to be used in the climate movement.

The first, hunger strikes, have long been used to compel the powerful to concede to demands for justice. Examples include a legend about Saint Patrick, Gandhi, the Suffragettes, farm labor leader Cesar Chavez, and members of the Irish Republican Army (IRA). Climate hunger strikers include a series of one-person actions in Europe and five young Americans who staged a two-week hunger strike outside the White House.

Hunger strikes have achieved mixed results. Legend has it that Saint Patrick won his negotiation with God, and history shows that public outrage over the forced feeding of hunger-striking Suffragettes helped advance voting rights for women. In Ireland, Prime Minister Margaret Thatcher refused to negotiate with hunger-striking Irish republican prisoners, ten of whom died in prison.

Should climate boomers stage hunger strikes? Although some of us have health conditions that would make this more dangerous for us than for younger, healthier people, we also have less to lose: most of our lives are behind us. Moreover, our status as elders could make a boomer hunger strike a shocking and therefore powerful tactic. I close that section by asking readers to imagine the potential impact of a nationwide boomer hunger strike for climate action. The challenge, I note, is to identify a demand that is worth the risk yet within reach.

I then turn to an even more taboo topic: using force to stop new fossil fuel infrastructure, describing two examples. The first involves two women who cut holes in empty sections of the Dakota Access Pipeline to delay construction. The second example involves a harrowing midnight attack on a construction site of a fracked gas pipeline in Canada. I explain how actions such as these can shift the Overton Window, for fellow activists and for powerful decisionmakers who benefit from the status quo.

I close the chapter by suggesting that climate boomers can boost the impact of radical actions with both audiences, by keeping an eye out for news reports of the next such action, learning and telling the stories of the brave people who undertake such acts, and by providing financial, moral and public support.

Three Final Bits of Advice

I close this book with three final bits of advice for new and emerging climate boomers, especially those who haven't interacted much with younger adults beyond their family circle.

First, be alert to changing social dynamics concerning race, gender, class and social privilege. Recognizing that climate change is fundamentally a question of social justice, climate activists place high value on diversity and inclusion, seeking out and lifting-up the voices of youth, members of frontline communities, queer people, women, and Black, Indigenous and People of Color (aka BIPOC).

If you have been out of the workplace for a while or are new to youth-led political movements, you may encounter social norms related to these values that are unfamiliar to you. Be prepared to listen and learn.

If you, like me, are an older white male, you will want to be especially aware of gender dynamics and read the room (whether virtual or in-person), when considering when to speak and what to say. Because we men—perhaps especially white men of a certain age—are accustomed to holding the floor, we sometimes need to step back to create space for others.

In the Th!rdAct Virginia coordinators group that I belong to, we track a queue or "stack" of those wanting to speak. If several people have indicated they would like to speak, white males drop to the bottom of the stack so that others can speak first. Asking participants to use the "raise hand" function, even in smaller meetings, is one way to create time and space for people who may not jump into a rough-and-tumble discussion.

You may be asked to "share your pronouns" when introducing yourself. This custom is grounded in the growing number of people, especially youth, whose gender identities are different than the sex assigned at birth or not readily apparent. Group leaders sometimes open meetings by stating their own pronouns ("Hi, I'm Sheila. My pronouns are she/her"), signaling others to do the same. In video calls, people may include their pronouns in their screen names. When you hear or see this happening, be prepared to follow suit.

You will also hear activists offer a land acknowledgement—a brief statement about Indigenous people's ties to the land where the speaker lives or where a meeting is being held. Land acknowledgements are grounded in recognition of the deep injustice of European settlement of Indigenous lands, including forced displacement and genocide against people who called the land home for many generations.

If this is new to you, take time to learn about the practice and write your own land acknowledgment. Native Land Digital (native-land.ca) is an online tool provided by an Indigenous-led Canadian NGO that offers "a platform where Indigenous communities can represent themselves and their histories on their own terms." Non-Indigenous people are invited to "learn more about the lands they inhabit, the history of those lands, and how to actively be part of a better future going forward together." The platform includes guidance on writing a land acknowledgment and a map where you can type in your address, discovering Indigenous people's ties to the land where you live or work.

Besides being aware of social conventions that may be new to you, consider how you present to a group—in the language of the young, how you "show up." If you normally mostly listen, consider speaking up more. If you normally talk a lot, practice holding back and listening. Either way, bring to your conversations an open, inquiring attitude.

My friend Ruth Levine, chief executive officer of IDInsight, a global advisory, data analytics, and research organization, recommends asking a simple question: "Can you tell me more?" This, she says, empowers the person who just spoke and leads them to explain things that might have otherwise been left unstated. I've tried it and it works.

Finally, volunteer for support tasks. In a virtual setting, this could include offering to take notes or schedule the next meeting. In a real-world setting, it may mean helping with food prep, clearing the table, or washing up. Such support is always welcome, regardless of who does it. Women—including many female boomers—have been socialized to take on such tasks. Male boomers are sometimes the least likely to think of it. Here's your extra nudge, guys!

My second piece of advice is much simpler: take stock of your digital toolkit and don't be afraid to ask for help in learning new computer skills. Most of us long ago mastered the digital basics like email, Facebook, and possibly Twitter. If your workplace used Microsoft tools like Outlook, take the time to learn the free Google tools commonly used in volunteer organizations, such as Google Docs and Google Groups. Most can be mastered quickly by watching free online videos.

Age need not be a barrier: Lani Ritter Hall, the 76-year-old Th!rdAct volunteer we met in Chapter 5, provides online Google Docs training for other Th!rdAct volunteers. "Some don't know what a menu bar is," she says. "Others just need a bit of help in setting up a new folder directory."

Two non-Google tools you may encounter are Proton Mail, a free encrypted alternative to Gmail, and Signal, an encrypted texting tool that lends itself to group threads. These are commonly used by activists organizing civil disobedience and other actions where keeping communications private is important. If you have a personal email account that is already overflowing with incoming messages, consider setting up a new account specifically for your climate volunteer work, either Gmail or Proton Mail. Both are free and easy to learn.

My final piece of advice concerns money. If you can afford to do so, front-load your climate donations, giving while you are still alive rather than waiting to give through your estate after you are gone. Because the window to avert catastrophe is closing fast, money given soon will be more impactful than money given many years from now. Also, because the climate movement badly needs a radical flank—and big institutional donors won't support one—your individual donations will go furthest if you give to small organizations and individuals who are helping to build one.

Money drives change and even a small share of boomer wealth, thoughtfully deployed, could make a big difference. What we don't spend ourselves will be passed on to our descendants and willed to charitable causes. According to a 2022 study, wealth transferred through 2045 will total $84.4 trillion—$72.6 trillion given to heirs and $11.9 trillion donated to charities.[66]

So far, shockingly little of this money is going to climate-related causes. Less than 2% of all charitable giving in 2021 went to address climate change. Foundations might be expected to do better, but a 2022 study found that, although nearly 90% of foundation leaders call climate change an urgent problem, only about 2% of money they pass out goes to climate-related projects, the same as for the philanthropic sector overall.[67]

Modest gifts can have an outsize impact when given to small organizations that are laser-focused on catalyzing system-wide change. This book mentions several; you will likely encounter more as you increase your engagement. Keep in mind that some worthy causes, like legal defense funds for people arrested for civil disobedience or sabotaging fossil fuel infrastructure, are not tax deductible. If you can afford to forego the tax deduction, giving money to such appeals is a powerful way to help save the planet.

If Not Now, When?

If you have read this far, I sincerely thank you. I know that some readers will disagree, perhaps strongly, with some of what I have said. Some will take exception to the idea of generational responsibility; after all, people hate being called to account, being "made to feel guilty."

Others may accept that our generation has an obligation to act, but balk at the idea that the time has come for mass civil disobedience, or they may be troubled by my call to support a radical flank. I would ask such readers to spend a day reading the latest climate news. Hardly a day passes without an alarming scientific report or a major climate-related disaster.

In just the few weeks since I began writing this final chapter, there have been unprecedented deluges in California, extraordinary record winter warmth across the northern hemisphere, record heat in Australia, and flooding in the Philippines that led the president to declare a "state of calamity." Scientists reported that Greenland's glaciers are melting faster than previously thought, while a separate sweeping study warned that half the glaciers outside of Greenland and Antarctica will vanish by the end of the century. New data showed that global and U.S. emissions continued to climb in 2022, with atmospheric concentrations of CO_2 passing 417 ppm, more than 50% above pre-industrial levels. The last time CO_2 concentrations were this high was more than 4 million years ago. Sea levels then were 15 to 75 feet higher than they are today, high enough to drown many of the world's largest cities.[68]

And yet, scientists assure us that a rapid end to the fossil fuel era can still make a big difference, slowing the rate of temperature increases and reducing the total amount that average temperatures will rise. Humans are causing climate change, and humans can still influence how it will unspool in the years ahead. Sadly, it's too late to restore the gentle, predictable climate that we knew in our youth. But it's not too late—and we are not too old—to save a livable planet for our children, grandchildren, and the generations coming after. Every tenth of a degree makes a difference.

Saving a livable planet will require that millions of us follow the path set forth in this book: start with meaningful personal actions, use your existing networks, join a climate focused organization, give time, give money, become a climate boomer. If you are willing and able, get trained and risk arrest in acts of non-violent civil disobedience. Support the emergence of a radical flank.

Of course, not everybody needs to do everything. In a 2022 article in *The Atlantic* titled "The Utility of White Hot Rage," author Emma Marris quotes advice from Mary Helgar, essayist and host of the climate podcast *Hot Takes*: "No matter what your current skills are, there's a way to use them to support climate justice," Helgar says. "Do what you are good at. If you can't do the work, care for people who can."

Her advice reminds me of two ancient Jewish teachings that are highly relevant to the situation we face today. The first states: "You are not obligated to complete the work, but neither are you free to desist from it." The second asks a question: "If not now, when?"

Chapter 9: Just Do It!

Action Checklist

- Take high-impact personal actions and spread the word
- Join a group and get connected to other activists
- Tap the power of organized religion
- Support non-violent civil disobedience and direct action
- Help build the radical flank

Endnotes

Chapter 1: Are Boomers to Blame?

1 Goodreads.com/quotes Retrieved February 2, 2023 at https://www.goodreads.com/quotes/547906-this-was-the-moment-when-the-rise-of-the-oceans

2 Remarks by the President at UN Climate Change Summit, September 23, 2014. ObamaWhiteHouse.archives.gov

3 Erickson, P. and Achakulwisut, P. (2021, June 24). *How Subsidies Aided the US Shale Oil and Gas Boom*. Stockholm Environmental Institute. Retrieved February 2, 2023, from https://www.sei.org/publications/subsidies-shale-oil-and-gas/

4 Burke's writings address this theme but there is no evidence that he wrote the quote widely attributed to him. Retrieved February 3, 2023, from https://tartarus.org/martin/essays/burkequote.html

5 *What's your birth carbon?* The Nature Conservancy. (n.d.). Retrieved February 3, 2023, from https://www.nature.org/en-us/get-involved/how-to-help/carbon-footprint-calculator/carbon-by-birth-year/

6 Hansen, J. (2021, December 20). *A Realistic Path to a Bright Future*. Retrieved February 19, 2023, from http://www.columbia.edu/~jeh1/mailings/2021/BrightFuture.03December2021.pdf

7 Buckley, Cara *'OK Doomer' and the Climate Advocates Who Say It's Not Too Late, The New York Times,* March 22, 2022, Updated March 26 2022, https://www.nytimes.com/2022/03/22/climate/climate-change-ok-doomer.html; see also Labayen, Evalena. *Environmental Activist Xiye Bastida Says "Ok, Doomers"* Interview Magazine, Dec. 2019, https://www.interviewmagazine.com/culture/environmental-activist-xiye-bastida-says-ok-doomers

Chapter 2: What Went Wrong?

8 Katznelson, Ira. *When Affirmative Action Was White: An Untold History of Racial Inequality in Twentieth-Century America*. W.W. Norton & Company (2006), (https://www.amazon.com/When-Affirmative-Action-White-Twentieth-Century/dp/0393328511)

9 Wikipedia. Global Climate Coalition, updated January 28, 2023, https://en.wikipedia.org/wiki/Global_Climate_Coalition#Opposition_to_Kyoto_Protocol

10 Cottle, Michelle, *The Disney for Boomers Puts Hedonism on Full Display, The New York Times*, Opinion, March 3, 2022

Chapter 3: Can Boomers Avert Catastrophe?

11 VisualCapitalist.com *Generational Power Index 2021*, https://www.visualcapitalist.com/wp-content/uploads/2021/05/generational-power-index-2021-1.pdf

12 Blazina, C., & DeSilver, D. (2023, January 31). *House gets younger, Senate gets older: A look at the age and generation of lawmakers in the 118th congress*. Pew Research Center. Retrieved February 3, 2023, from https://www.pewresearch.org/fact-tank/2023/01/30/house-gets-younger-senate-gets-older-a-look-at-the-age-and-generation-of-lawmakers-in-the-118th-congress/

13 Pew Research Center, *Key findings: How Americans' attitudes about climate change differ by generation, party and other factors* (Published June 2, 2021, Accessed February 2, 2023), https://www.pewresearch.org/fact-tank/2021/05/26/key-findings-how-americans-attitudes-about-climate-change-differ-by-generation-party-and-other-factors/

14 *Do younger generations care more about global warming? Climate Change in the American Mind*. Published June 12, 2019, Accessed February 22, 2023, https://climatecommunication.yale.edu/publications/do-younger-generations-care-more-about-global-warming/

15 *Climate change and civil disobedience. Climate Change in the American Mind.* (Published February 2022. Retrieved February 2, 2023, https://climatecommunication.yale.edu/publications/who-is-willing-to-participate-in-non-violent-civil-disobedience-for-the-climate/

16 Nadeem, R. (2022, April 28). *Gen Z, millennials stand out for climate change activism, social media engagement with issue.* Pew Research Center. Retrieved February 2, 2023, from https://www.pewresearch.org/science/2021/05/26/gen-z-millennials-stand-out-for-climate-change-activism-social-media-engagement-with-issue/

17 Moody, Harry R. "Baby Boomers: From Great Expectations to a Crisis of Meaning." Generations: Journal of the American Society on Aging, vol. 41, no. 2, 2017, pp. 95–100. JSTOR, https://www.jstor.org/stable/26556289. Accessed 2 Feb. 2023.

Chapter 4: Every Little Bit Helps, Right?

18 Dunaway, F. (2015) *Seeing Green: The Use and Abuse of American Environmental Images,* The University of Chicago Press (https://press.uchicago.edu/ucp/books/book/chicago/S/bo13666193.html)

19 Kaufman, M. (2021, July 9). "The Devious Fossil Fuel Propaganda We All Use." *Mashable*. Retrieved February 2, 2023, from https://mashable.com/feature/carbon-footprint-pr-campaign-sham

20 Wagner, G. (2012). *But Will the Planet Notice? How Smart Economics Can Save the World.* Hill and Wang.

21 Wagner, G. (2021, August 12). *How I greened my prewar co-op*. New York Magazine, Retrieved February 2, 2023, from https://gwagner.com/curbed-coop/

22 Bollinger, B., Gillingham, K., Kirkpatrick, A. J., & Sexton, S. (2019, June 24). *Visibility and peer influence in durable good adoption*. SSRN. Retrieved February 2, 2023, from https://papers.ssrn.com/sol3/papers.cfm?abstract_id=3409420

23 Searchinger, T., Waite, R., Hanson, C., Ranganathan, J., & Matthews, E. (2020, July 7). *Creating a Sustainable Food Future.* World Resources Institute. Retrieved February 2, 2023, from https://www.wri.org/research/creating-sustainable-food-future

24 Ritchie, H. (2020, January 24). "You want to reduce the carbon footprint of your food? Focus on what you eat, not whether your food is local." *Our World in Data.* Retrieved February 3, 2023, from https://ourworldindata.org/food-choice-vs-eating-local#:~:text=Overall%2C%20animal%2Dbased%20foods%20tend,CO2%2Dequivalents%2C%20respectively.

25 *Walking for fitness*. Mayo Clinic Health System. (2022, October 4) Retrieved February 2, 2023, from https://www.mayoclinichealthsystem.org/hometown-health/speaking-of-health/walking-for-fitness

26 *Public Transportation Reduces Greenhouse Gasses and Conserves Energy*. American Public Transportation Association (2008)

27 *Banking on Climate Chaos 2022* https://www.bankingonclimatechaos.org/

28 Hicks, C. (2021, November 4). *9 ESG Tools for Sustainable Investors*. Kiplinger.com. Retrieved February 2, 2023, from https://www.kiplinger.com/investing/esg/603706/esg-tools-for-sustainable-investors

29 The United States Government. (2021, August 5). *Fact sheet: President Biden announces steps to drive American leadership forward on clean cars and trucks*. The White House. Retrieved February 2, 2023

30 Wille, M. (2022, June 14). *BMW is testing an IX SUV with a driving range of 600 miles*. Input. Retrieved February 2, 2023, from https://www.inputmag.com/tech/bmw-ix-suv-battery-our-next-energy-600-miles-range

31 Reid, C. (2022, October 12). *Electric car batteries lasting longer than predicted delays recycling programs*. Retrieved February 2, 2023, from https://www.forbes.com

32 *Community Solar Resources*. Energy.gov. Retrieved February 2, 2023, from https://www.energy.gov/communitysolar/community-solar-resources

33 Husabø, I. A. (2020, November 19). *1% of People Cause Half of Global Aviation Emissions. Most People in Fact Never Fly.* https://partner.sciencenorway.no/. Retrieved February 2, 2023, from https://partner.sciencenorway.no/climate-change-global-warming-transport/1-of-people-cause-half-of-global-aviation-emissions-most-people-in-fact-never-fly/1773607

34 Hanes, S. (2022, April 18). "Grounded, and Loving It. Can Giving Up Air Travel Bring Joy?" *The Christian Science Monitor*. Retrieved February 2, 2023, from https://www.csmonitor.com/Environment/2022/0418/Grounded-and-loving-it.-Can-giving-up-air-travel-bring-joy

35 *Carbon offsets*. Sustainable Travel International. (2023, January 19). Retrieved February 4, 2023, from https://sustainabletravel.org/our-work/carbon-offsets/

36 Monbiot, G. (2006, October 18). *Paying for Our Sins*. The Guardian. London

37 CAPA — Centre for Aviation. (2021, March 21). *United's Kirby: Carbon Offsets "a fig leaf for a CEO to write a check"*. Retrieved February 4, 2023, from https://centreforaviation.com/analysis/reports/uniteds-kirby-carbon-offsets-a-fig-leaf-for-a-ceo-to-write-a-check-555398

38 Niina H. Farah, L. C. (2022, November 29). "EPA Floats Sharply Increased Social Cost of Carbon." *E&E News*. Retrieved February 4, 2023, from https://www.eenews.net/articles/epa-floats-sharply-increased-social-cost-of-carbon/#:~:text=That%20document%20estimates%20the%20social,future%20impacts%20of%20climate%20change

39 Berger, Sebastian (2022, January 24). *Willingness-to-Pay for Carbon Dioxide Offsets: Field Evidence on Revealed Preferences in the Aviation Industry.* Global Environmental

Change. Retrieved February 2, 2023, from https://www.sciencedirect.com/science/article/pii/S0959378022000085

Chapter 5: How Can I Work with Others?

40 MarcusBuckingham.com *The Gift of Standout* Retrieved February 2, 2023, from https://www.marcusbuckingham.com/gift-of-standout/

41 Joselow, M., & Montalbano, V. (2022, March 10). "Heat pumps can counter Putin and the Climate Crisis, advocates say. The White House is listening." *The Washington Post*. Retrieved February 2, 2023, from https://www.washingtonpost.com/politics/2022/03/10/heat-pumps-can-counter-putin-climate-crisis-advocates-say/

42 *President Biden invokes defense production act to accelerate domestic manufacturing of Clean Energy*. Energy.gov. Retrieved February 2, 2023, from https://www.energy.gov/articles/president-biden-invokes-defense-production-act-accelerate-domestic-manufacturing-clean

43 MacDonald, Lawrence and Cao, Jing (2014). *The Sudden Rise of Carbon Taxes 2010–2030*. Center for Global Development. Retrieved February 2, 2023, from https://www.cgdev.org/publication/ft/sudden-rise-carbon-taxes-2010-2030

44 *The Carbon Border Adjustment Mechanism (CBAM) – the European Union's next climate action step*. JD Supra. (2022, December 22). Retrieved February 2, 2023, from https://www.jdsupra.com/legalnews/the-carbon-border-adjustment-mechanism-1730794/

45 The Years Project. (2022, June 19). *Fighting Polluting Industry in Cancer Alley*. YouTube. Retrieved February 4, 2023, from https://www.youtube.com/watch?v=XjwpfvPdRcw

46 *New Case Study Details Hampshire College's Mission-based Endowment Investing.* Hampshire College. (2017, November 17). Retrieved February 2, 2023, from https://www.hampshire.edu/news/new-case-study-details-hampshire-colleges-

mission-based-endowment-investing#:~:text=Hampshire%20is%20widely%20known%20for,to%20divest%20from%20fossil%20fuels

47 Fossil Fuel Commitments Database (2021). Retrieved February 2, 2023, from https://divestmentdatabase.org/

48 Spence, R. (2021, November 24). *Harvard Gives In to Pressures to Divest from Fossil Fuels (Mostly).* Corporate Knights. Retrieved February 4, 2023, from https://www.corporateknights.com/issues/2021-11-education-and-youth-issue/harvard-divests/

49 Divest Harvard. (2021). *Beyond the Endowment: Uncovering Fossil Fuel Interests on Campus.* Retrieved February 5, 2023, from https://www.divestharvard.com/wp-content/uploads/2021/11/BeyondTheEndowmentFFDH.pdf

50 *Public Pension Funds Continue to Invest Billions into Fossil Fuels.* Climate Safe Pensions. (2022, November 17). Retrieved February 5, 2023, from https://climatesafepensions.org/

Chapter 6: What's Faith Got to Do with It?

51 Sena, K. (2020, October 22). *Recognizing Indigenous People's Land Interests Is Critical for People and Nature.* World Wildlife Fund. Retrieved February 5, 2023, from https://www.worldwildlife.org/stories/recognizing-indigenous-peoples-land-interests-is-critical-for-people-and-nature#:~:text=By%20fighting%20for%20their%20lands,they%20have%20lived%20for%20centuries.

52 Estimates of Americans' religious affiliations differ by source and change over time. For this account I consulted several sources, including the Public Religion Research Institute (PRRI) *2020 Census of American Religion* (2021, July 8). Retrieved February 2, 2023, from https://www.prri.org/press-release/prri-releases-groundbreaking-2020-census-of-american-religion/)

53 Carter, M. (2013, July 1). *United Church of Christ to Become First U.S. Denomination to Move Toward Divestment from Fossil Fuel Companies.* United Church of Christ. Retrieved February 2, 2023, from https://www.ucc.org/gs2013-fossil-fuel-divestment-vote/

54 Pew Research Center. (2022, June 13). *Religious Landscape Study.* Pew Research Center's Religion & Public Life Project. Retrieved February 5, 2023, from https://www.pewresearch.org/religion/religious-landscape-study/compare/immigrant-status/by/religious-tradition/

55 Mitchell, T. (2022, October 6). *1. The size of the U.S. Jewish population.* Pew Research Center's Religion & Public Life Project. Retrieved February 6, 2023, from https://www.pewresearch.org/religion/2021/05/11/the-size-of-the-u-s-jewish-population/

Chapter 7: Should I Get Arrested?

56 *Climate Change and Civil Disobedience. Climate Change in the American Mind.* (Published February 2022. Retrieved February 2, 2023, https://climatecommunication.yale.edu/publications/who-is-willing-to-participate-in-non-violent-civil-disobedience-for-the-climate/

57 *Global Warming's Six Americas.* (2022, February 10). Retrieved February 2, 2023, from https://climatecommunication.yale.edu/about/projects/global-warmings-six-americas/

58 *Why Civil Resistance Works.* Columbia University Press. (2012). Retrieved February 2, 2023, from http://cup.columbia.edu/book/why-civil-resistance-works/9780231156820

59 Wikimedia Foundation. (2022, October 7). *Raging grannies.* Wikipedia. Retrieved February 2, 2023, from https://en.wikipedia.org/wiki/Raging_Grannies

60 *History of the 1,000 Grandmothers.* 1000 Grandmothers. (n.d.). Retrieved February 2, 2023, from http://www.1000grandmothers.com/our-history.html

61 YouTube. (2022, May 19). *Peter Kalmus on Courage | Extinction Rebellion UK*. Retrieved February 2, 2023, from https://www.youtube.com/watch?v=NH1m7DYUj08

62 Brown, M. (2021, June 9). *Keystone XL pipeline nixed after Biden stands firm on permit*. AP NEWS. Retrieved February 6, 2023, from https://apnews.com/article/donald-trump-joe-biden-keystone-pipeline-canada-environment-and-nature-141eabd7cca6449dfbd2dab8165812f2

Chapter 8: Who You Callin' Radical?

63 Fukuyama, F. (2022, June 16). *We're Cooked.* American Purpose. Retrieved February 2, 2023, from https://www.americanpurpose.com/blog/fukuyama/were-cooked/

64 "Protecting Water Is Never Terrorism: Repeal Jessica Reznicek's Terrorist Enhancement!" The Action Network. Retrieved February 2, 2023, from https://actionnetwork.org/petitions/protecting-water-is-never-terrorism-repeal-jessica-rezniceks-terrorist-enhancement/thankyou

65 Schroder, A. (2022, January 24). "Fight Like an Animal: The Life and Death of Radical Environmentalism." *Apple Podcasts*. Retrieved February 2, 2023, from https://podcasts.apple.com/ie/podcast/the-life-and-death-of-radical-environmentalism/id1511430305?i=1000548911633

Chapter 9: Just Do It!

66 *Press release: Cerulli anticipates $84 trillion in wealth transfers through 2045.* Cerulli Associates. (2022, January 20). Retrieved February 2, 2023, from https://www.cerulli.com/press-releases/cerulli-anticipates-84-trillion-in-wealth-transfers-through-2045

67 *Much Alarm, Less Action: Foundations & Climate Change.* The Center for Effective Philanthropy. (2022, January 14). Retrieved February 2, 2023, from https://cep.org/event/

july-14-2022-much-alarm-less-action-foundations-climate-change/

68 *Carbon Dioxide Now More Than 50% Higher Than Pre-Industrial Levels.* National Oceanic and Atmospheric Administration. (2022, June 3). Retrieved February 23, 2023, from https://www.noaa.gov/news-release/carbon-dioxide-now-more-than-50-higher-than-pre-industrial-levels

Transform your life, transform *our* world. Changemakers Books publishes books for people who seek to become positive, powerful agents of change. These books inform, inspire, and provide practical wisdom and skills to empower us to write the next chapter of humanity's future.
www.changemakers-books.com

The Resilience Series

The Resilience Series is a collaborative effort by the authors of Changemakers Books in response to the 2020 coronavirus pandemic. Each concise volume offers expert advice and practical exercises for mastering specific skills and abilities. Our intention is that by strengthening your resilience, you can better survive and even thrive in a time of crisis.
www.resiliencebooks.com

Adapt and Plan for the New Abnormal – in the COVID-19 Coronavirus Pandemic
Gleb Tsipursky

Aging with Vision, Hope and Courage in a Time of Crisis
John C. Robinson

Connecting with Nature in a Time of Crisis
Melanie Choukas-Bradley

Going Within in a Time of Crisis
P. T. Mistlberger

Grow Stronger in a Time of Crisis
Linda Ferguson

Handling Anxiety in a Time of Crisis
George Hoffman

Navigating Loss in a Time of Crisis
Jules De Vitto

The Life-Saving Skill of Story
Michelle Auerbach

Virtual Teams – Holding the Center When You Can't Meet Face-to-Face
Carlos Valdes-Dapena

Virtually Speaking – Communicating at a Distance
Tim Ward and Teresa Erickson

Current Bestsellers from Changemakers Books

Pro Truth

A Practical Plan for Putting Truth Back into Politics

Gleb Tsipursky and Tim Ward

How can we turn back the tide of post-truth politics, fake news, and misinformation that is damaging our democracy? In the lead-up to the 2020 US Presidential Election, *Pro Truth* provides the answers.

An Antidote to Violence

Evaluating the Evidence

Barry Spivack and Patricia Anne Saunders

It's widely accepted that Transcendental Meditation can create peace for the individual, but can it create peace in society as a whole? And if it can, what could possibly be the mechanism?

Finding Solace at Theodore Roosevelt Island

Melanie Choukas-Bradley

A woman seeks solace on an urban island paradise in Washington D.C. through 2016–17, and the shock of the Trump election.

the bottom

a theopoetic of the streets

Charles Lattimore Howard

An exploration of homelessness fusing theology, jazz-verse and intimate storytelling into a challenging, raw and beautiful tale.

The Soul of Activism
A Spirituality for Social Change
Shmuly Yanklowitz
A unique examination of the power of interfaith spirituality to fuel the fires of progressive activism.

Future Consciousness
The Path to Purposeful Evolution
Thomas Lombardo
An empowering evolutionary vision of wisdom and the human mind to guide us in creating a positive future.

Preparing for a World that Doesn't Exist – Yet
Rick Smyre and Neil Richardson
This book is about an emerging Second Enlightenment and the capacities you will need to achieve success in this new, fastevolving world.